...And Nothing Happened --

But You

Can Make It Happen!

Medical and health technologies that are safer and more effective than those in current use have been bypassed in our medical system. Have greed, ego and vested interests kept these incredible technologies from you?

By
Steven A. Ross

Steven A Ross
PO Box 20756
Sedona, AZ 86341
www.andnothinghappened.com

Cover illustration and design by Tony De Luz.
www.tonydeluz.com

Disclaimer

The following disclaimer is required due to certain vested interests that exist within the health field. As is pointed out in the article "The Tomato Effect," there are entities that don't like competitive products or technologies that are non-pharmaceutical.

Disclaimer: The information contained within this book is not intended as a diagnosis or recommendation of therapy for any health condition that the reader might have. You should always consult a health professional if you have a question. Although there are various machines that might be using the technologies mentioned in this book, this is not an endorsement of any product existing on the market.

Note: Apparently, there is no problem for you to take the pharmaceutical products that are recommended to you through the advertising media. Remember some of their side effects are *minimal* and might include heart attack, stroke, coma or other *minor* problems such as personality changes and a desire to commit suicide. Please read product labels carefully and remember that you should consult the Physician's Desk Reference to learn of other adverse effects not listed on the bottle or possible problems your medical specialist has not warned you about.

ISBN 978-0-578-01687-0

"WHEN WE UTTER THE WORDS THAT

OUR CONDITION IS HOPELESS OR INCURABLE,

WE DENY THE UNLIMITED BOUNTIES THAT EXIST AROUND US.

OUR CREATOR HAS ENTRUSTED NATURE TO PROVIDE US

WITH FOUNTS OF ENDLESS ENERGY

THAT CAN BE TAPPED

THROUGH THE INGENUITY OF THE HUMAN MIND AND WILL."

Steven A. Ross

Acknowledgements

I express my gratitude to Edward Monroe who started me on my lifelong quest to learn more about my purpose in life as well as my place in the universe. Although he is no longer in the physical, his influence continues to guide my journey of continuous discovery.

My appreciation and admiration go out to Robert Maver for the courage he showed in establishing a worldwide research division through Mutual Benefit Life Insurance Company and for his believing that the World Research Foundation would justify his faith that there are many efficacious therapeutic techniques beyond what his allopathic experts had been reporting to him.

I want to thank Cynthia Richmond for her support and to John Wade for his time and energy to do proof reading and provide many helpful suggestions to make this book more user-friendly.

I would also like to thank LaVerne Boeckmann who co-founded the World Research Foundation and has worked tirelessly to keep the foundation alive and well. The World Research Foundation has provided the resources that have allowed me to live the life that most people only dream about living.

I was fortunate to have had wonderful parents. Adele and Stan were my role models throughout my life. They demonstrated through their actions what a loving marital relationship could be. There is no way that I could ever adequately express my appreciation to them. Special loving thanks to my Mom who has always been the first person to support my dreams and encourage me to carry forward when there didn't appear to be a pathway.

It is with my deepest love and gratitude that I thank Prof. Dr. Karl Walter and Heidi Kleber for their more than 24-year friendship and their incredible scheduling that allowed me to make important discoveries throughout Europe.

To Adriana, my thanks for your support and insistence that this material be presented to the public sooner than later – the "sooner" meaning now, and that is why you are reading this book.

I would like to thank the philosophers, poets, writers and musicians who have brought such incredible beauty into my life. It is in the quest of the One, the Good and the Beautiful that I have dedicated my life and energy.

It is in the quest for the beautiful that this book is dedicated.

Table of Contents

Preface..7
    Why I wrote this book
Introduction
    Part 1 The Emperor's New Clothes.................................12
      Analogy of what is taking place in medicine
    Part 2 The Tomato Effect.......................................18
      Why new research is often not accepted
    Part 3 How Do We Treat The Messengers – Past, Present and Future........22
      What happens to the discovers of new therapies
Chapter 1 Preparing For My Journey...............................26
    Brief overview of experiences leading to writing this book
Chapter 2 Surpassing All Limitations.............................30
    Story of Royal Rife who invented the world's most powerful
    microscope, ten times more powerful then those used today.
    The microscope was developed in the 1920s.
Chapter 3 The Nobel Chairman Is Ignored..........................56
    Dr. Bjorn Nordenstrom, a past Chairman of the Nobel Assembly,
    discovered how to shrink lung and breast cancer tumors using
    electricity with no side effects. His work was completely ignored
Chapter 4 Who Is Colorblind?.....................................65
    The use of color for various diseases and illnesses. Doctors swore
    under oath in court they had healed cancer, arthritis, diabetes and
    cured a 8 year old girl of 3rd degree burns without topical therapy.
    This color system was used in the 1920s.
Chapter 5 Waves That Heal..81
    Georges Lakhovsky discovered the electrical properties of cells
    and built a device called a MWO destroying cancer and accelerating
    healing. The device was built in the 1930s.
Chapter 6 Fields of Life...88
    Dr. Harold Burr, Yale University, documents the invisible energy
    fields surrounding all living organisms. These L-fields keep all matter
    in shape. Burr could predict where and when a cancer tumor would
    develop before it was ever seen or registering on any medical test
    equipment. Research was done in the 1930s.
Chapter 7 The Phenomenon of Life.................................95
    Dr. George Crile discovered that electricity is the key to growth and
    health of all living organisms. Recognized as one of the top ten
    surgeons of all time, he could explain why and how life functions and
    why illness develops. Time period was the 1930s.
Chapter 8 Twice Normal Speed....................................100
    This is the story of the amazing Diapulse Machine. It was confirmed
    by more than twenty universities and research centers to accelerate healing
    twice the normal speed. Used in the Olympics, the FDA banned it for 15
    years because they claimed it couldn't possibly work. Hundreds of
    double-blind studies all around the world confirm it works.

Chapter 9   Bypassing Bypass...107
In the 1940s, a doctor discovered a little herb from Africa that balances
out the pH in the heart and eliminates the need for almost 90% of scheduled
bypass surgeries. Hundreds of studies including one from a Nobel Prize
winner, confirms that the model used for heart and heart bypass therapy is not
accurate or valid.

Chapter 10   Visualizing The Meridians...116
Two aerospace nuclear medical doctors have made visible the entire
acupuncture meridian system using radioactive isotope (Te99) and catscan
cameras. They have validated the ancient Chinese meridian system as
being 100% correct as written more than 3,000 years ago.

Chapter 11   The Cast Off Cell..120
A presentation made in 1954 to the Gynecological Society in Chicago,
Illinois, reported a new test that measured the voltage of one cell cast off
in the vaginal tract that could determine whether a woman had uterine,
cervical or no cancer. There was never a false negative and the positive
was 94% accurate. It was a completely non-invasive accurate test.

Chapter 12   Sam the Eagle...123
Numerous veterinarians tried to heal the injured wing of Sam the Eagle.
When all efforts failed, a special electromagnetic healing device was used.
This European machine healed the eagle's wing but placed the doctor at
great risk from medical authorities in the U.S.

Chapter 13   The Cure That Time Forgot...126
Ultraviolet blood irradiation was used in the 1940s and had a cure rate of
98% in early and moderately advanced infections and 50% with near dead
patients. It was used for wounds, asthma and arthritis. With the advent of
pharmaceuticals all use of this technique disappeared. There were never
any reported side effects.

Chapter 14   Have We Left A Stone Unturned?..131
The vice-president and head actuary for the 13th largest insurance
company in the US gives his opinion on what is taking place in the American
health-care system.

Chapter 15   Conclusion..142
Chapter 16   And What Must Happen..147
What must happen in the U.S. for there to be a change in our medical
approach for the leading causes of illness and death. The preface and
introduction explained how the U.S. is at the lowest rank of the major
industrialized countries regarding mortality rates.

Appendix...149
    Rife Microscope: Letters and newspaper articles.................................150
    Time Capsule: Short articles on health from magazines 1930-1952.................159
    The Cell: A simple electrical model of a nominal biological cell................171
Suggested Reading...172

Preface

I have always been a person who loves mysteries. Looking back over my life my greatest satisfaction has come in solving something difficult to understand or explain. Often times something that I considered a mystery might be very well-known and understood by other people, but for me it offered an opportunity to learn.

One of the major mysteries of my life that would lead to my professional activities for the last 40 years had to do with a health situation that I faced.

Back in the late 1960s, I was on an athletic scholarship at a university in the Los Angeles area. I ran 100 meters, 200 meters and anchored our 400 meter relay team. I was very blessed in my life with having beyond average speed and held all of the school records for my event at my high school and at my university Cal-State Northridge.

During the Easter break all the track athletes would work out on their own schedules. During one of the days I was working out, I severely injured my left knee. I was traveling at about 80 percent of my top speed when I tried to extend my stride to avoid a sprinkler head that had been left on the running course. I immediately went to the ground in tremendous pain. Although the trainer's room was open at the time there was no other athlete around me, and I had to crawl across the campus on my hands and knees to get some help. Somehow I knew I had torn the tendons and ligaments in my knee.

My coach was called at his home, and he rushed down to see what had happened to me. Within days, I was sent to one of the leading sports physicians in the United States. This physician was consulted by the then Los Angeles Rams, Lakers and Dodgers. In fact at that time most athletic teams sent their athletes to this doctor.

After his diagnosis, he told me that I needed to have surgery or I could never compete again. For a second opinion, the university sent me to one of the premier trainers in the country who was working at UCLA. His name was Ducky Drake, and he also told me that I would need to have surgery.

I was extremely dejected at that moment. I played many sports throughout my youth and always felt that nothing was impossible for me. I remember, even in elementary school, loving to run across the playground to see how fast I could run. I was always ahead of my age group with my speed.

I returned to my trainer's room and noticed a periodical next to one of the whirlpools. It was a magazine with mechanical information dealing with engines and machines, and there was an article that seemed out of place. The article dealt with the effects of cold temperatures upon different mechanical operations. It also explored the use of cold therapy for injuries. I became intrigued with the information and wondered if there might be any possibility that I could benefit from something like that.

I placed a call to the sports physician who gave me a very blunt and short answer. The therapy was absolute trash and would not work. In fact, he mentioned that it was some type of alternative therapy that was extremely questionable. I asked the doctor what he knew about such cold therapy coming out of Eastern Europe. He said he did not know anything about it. The physician's final statement was, "Son, this is America, and if we don't use it it has no validity!" I was 19 years old at the time and could not understand how someone could tell me something was worthless when they did not know anything about it and did not want to learn anything about it.

I decided to utilize the treatment on my own. During the next three weeks, I went into the trainer's room, went to the ice machine, rolled the ice into a ball, and massaged my knee with the ice until it was numb. I would sit and wait for the knee to return to room temperature, and then I would ice it again. I was relentless and did this all day long. It obviously was not a technically oriented therapy and I continued the regime despite the doctor's belief that it would be useless for my situation.

Three weeks later, I resumed limited training, a reduced schedule, and at the end of that season, I anchored my university relay team to place in the top five at the small college Track and Field Nationals. I earned my All-American status that year. The 100 meter times I ran throughout the season would have placed me in the top five in the 100 meter national finals but there were too many preliminary races and I was not able to make it to the finals that year.

Anyone with common sense knows that you cannot fake your way to the finals in track. If my knee had not been restored to healthy function, I could not have competed and placed at the national level as I did.

Perhaps what's more important, a seed was planted for me. Why was it that the so-called experts did not know about this technique and what other things might exist around the world that we were not familiar with at the time?

What starts as an interest often becomes a full profession. I began making contacts and gathering information regarding therapies, techniques and clinics around the world. In 1984, I met LaVerne Boeckmann who had the same interests in health and who had gone through an incredible journey of her own, locating a therapy that helped her with a specific medical condition. Because of somewhat similar situations we decided to co-found an organization called the World Research Foundation, a non-profit organization whose purpose was to gather information from all over the world on every therapy for every disease and illness. The gathered information encompasses health data from ancient times to the present and from allopathic to alternative sources.

I have been gathering information pertaining to health from 1969 to the present. What I have discovered has created both happiness and consternation. Happiness in the fact that there are solutions to many of the health problems that I have heard in the U.S. are incurable and at the same time anger that techniques are being used outside the U.S. not available in our country.

I have pondered the fact that in one place in the world certain medical conditions are addressed with successful therapies that are not available or banned in another part of the world.

My searching has consisted of making contacts around the world and then actually visiting clinics, hospitals, therapists, medical doctors and scientists in their laboratories or research centers, seeing for myself what research is taking place and the manner in which the researchers handle themselves. My journey has taken me to many countries in Europe, Asia, South America and the Far East.

During the years, I have been fortunate to achieve a reputation that has been in demand for radio, television and the print media. I have been a guest on hundreds of radio programs and conducted lectures all over the world. This includes presentations to members of Parliament in the United Kingdom, science members of the Latvian Academy of Science, as well as appearing on Latvian television; presentation at the Chinese Academy of Medical Science in Beijing and other foreign countries.

I have acted as a consultant for the City of Los Angeles in its dispute with the State of California over aerial spraying of pesticides and was credited in several sources as providing the information that made the difference in the decision of council members in Orange County, California.

Furthermore, I have presented lectures for continuing educational credits for medical professionals and lectured at medical hospitals and for medical networks. For 2 ½ years I was the head researcher and worldwide consultant for Mutual Benefit Life Insurance Company.

I don't mean to be tooting my own horn, but I have seen many things that the average people or medical doctors will never see during their lives. Back in 1978 there was an article in a medical magazine, Annals of Internal Medicine, that mentioned that in order for a biomedical expert to stay current in their field, he would have to read 18,000 articles per week. I used to make the statement that even if that expert was reading away; he would fall behind if he ever took a vacation. However, it is not just enough to read articles or listen to a short synopsis of someone's opinion. It makes a difference when you see with your own eyes and ask questions of researchers.

I had the opportunity one year to visit some of the top medical clinics in Europe. I have shared information about them in this book. I believe that you will be overwhelmed by the manner in which these clinics operate. I might add that they are as large physically as the biggest hospitals in the U.S.. They use techniques that should have been incorporated in our hospitals a long time ago. But we seem to have an arrogance much like the doctor who treated me during my university days.

I love my country very much but the U.S. has been letting me down when it comes to health. Why is it that cancer, multiple sclerosis, diabetes, arthritis and heart problems are addressed in a different, yet more successful manner, in other parts of the world? Why is there a group of people here who have appointed themselves the guardian of truth, the quack busters, who seem to discredit everything that is not pharmaceutical? Techniques that are successfully used in Europe and Asia are considered in this country questionable and with no validity, yet European and Asian cultures have histories many thousands of years old, the U.S. is the newcomer to the health care arena.

I, for one, am tired of seeing so-called guardians of the truth and listening to their drivel of half-truths and information bordering on lies. If someone makes the comment that there is very little scientific data and at the same time does not allow testing to acquire that data, that is unfair and malicious. With so many people dropping dead in clinical trials of a new diabetes drug, as was reported at the time of this writing, why aren't these quack busters looking into these questionable therapies? In this book, I am sharing some of the most important therapies and techniques that I have discovered during the years. These are discoveries that affect everyone in the world.

I am not pushing these therapies to make personal money from them or from someone I know who is involved with their use. I am writing this book because I believe without health you have NOTHING. You can be the most beautiful looking person, have great power, have tremendous wealth, but if you don't have health, what does it matter?

Leading medical and health specialists in the world have made the discoveries I've written about in this book - a former Chairman of the Nobel Assembly, a professor emeritus at Yale University, the top researchers from the Mayo Clinic, Johns Hopkins, Northwestern University and other individuals who have been affiliated with the most famous research centers and hospitals in the world.

They presented their discoveries to the world - *And Nothing Happened*!

BUT YOU CAN MAKE IT HAPPEN!

Introduction
Part 1
The Emperor's New Clothes: 2008

"Then the emperor walked at the head of the procession under the beautiful canopy, and everyone in the streets cried, 'look at the emperor's new clothes. Are they not the most wonderful he has ever worn?' They did not dare admit they could see nothing for

Vilhelm Pedersen Illustrator

fear they would be called fools. Never before had the emperor's clothes been so much admired. 'But he has got nothing on', said a little child. 'Oh, listen to the innocent,' said the father. And one person whispered to another what the child had said. 'He has nothing on. A child says he has nothing on!' 'But he has nothing on,' all the people cried at last! The emperor felt a shudder go through him, for he knew at once that it was true, but he had to lead the procession." - Hans Christian Anderson.

There is something very wrong with our medical health care system and medical therapeutics. There is something wrong with pharmaceutical products that have side effects of heart attack, stroke, memory loss, change of personality, liver damage, kidney damage, blindness, desire for gambling, and on and on and on. What happened to the adage, associated with the Hippocratic Oath taken by every licensed physician, "First do no harm"?

- Doctors are the third leading cause of death in the U.S. behind cancer and heart disease, causing 250,000 deaths every year. [1]
- The University of Toronto did a study and found that roughly 2,216,000 patients in U.S. hospitals per year experienced a serious adverse drug reaction during their hospital stays. Based on these figures, adverse drug reactions are now the fourth leading cause of death in this country. [2]
- The number of people having in-hospital, adverse drug reactions (ADR) to prescribed medicine is 2.2 million. Dr.

[1] Journal of the American Medical Association; Vol. 284, July 26, 2000
[2] JAMA 98;279 (15): 1200-5

Richard of the CDC, in 1995 said the number of unnecessary antibiotics prescribed annually for viral infections was 20 million. Dr. Besser, in 2003 now refers to tens of millions of unnecessary antibiotics.

- Dr. Howard Brody, M.D., director of the Institute for the Medical Humanities at the Univ. of Texas Medical Branch points to a national survey published in 2007 in the New England Journal of Medicine, in which 94% of the doctors polled said they had "direct ties" to the drug industry. [3]
- Pharmaceutical industry spending on marketing to doctors has risen 275 percent from 1996 to 2004. This is $7 billion a year on perks and $18 billion on free samples. [4]
- Americans today take more medications than in any other country in the world. Spending on direct-to-consumer drug advertising has increased more than 300% in nearly a decade to $4.2 billion, up from $1.1 billion in 1997.
- France is the healthcare leader, U.S. comes dead last. The study conducted by the Commonwealth fund published in January, 2008, showed the United States dead last of 19 industrialized countries. All countries have improved substantially except for the U.S., said Ellen Nolte, the lead author of the study. [5]
- "The huge sums now spent in the name of medical progress produce only marginal improvements in health. America devotes nearly 12% of is GNP to its high-technology medicine, more than any other developed country. Yet, overall, Americans die younger, lose more babies, and are at least as likely to suffer chronic diseases." [6]

In a recent comparison of 13 industrialized countries, the United States ranked an average of 12th in the following statistics as reported by the World Health Organization Report, 2000. The statistics refer to the percentage of individuals who will live to these ages. [7]

- 13th for neonatal mortality and infant mortality

[3] AARP, Vol. 49. No.1, January-February 2008, pg.20.
[4] ibid
[5] www.commonwealthfund.org/publications/
[6] The Economist, October 20, 1990
[7] World Health Organization, World Report 2000

- 11th for post neonatal mortality
- 10th for life expectancy at age 15 years for females, 12th for males
- 10th for life expectancy at age 40 years for females, 9th for males
- 7th for life expectancy at age 65 years for females, 7th for males
- 10th for age-adjusted mortality.

The Economist also reported the following:
- According to a 1988 study conducted in Europe, coronary by-pass surgery is beneficial only in the short term. A bypass patient who dies within five years would probably have lasted longer if he had simply taken drugs.
- An American study completed in 1988 concluded that removing tissue from the prostate gland after the appearance of (non-cancerous) growths, but before the growths can do much damage, does not prolong life expectancy. Yet the operation was performed regularly and cost Medicare, the federally-subsidized system for the elderly, more than $1 billion a year.
- Dr. Louise Russell, a professor of economics at Rutgers University in New Jersey, reports that although anti-cholesterol drugs have been shown in clinical trials to reduce the incidence of death due to coronary heart disease, in ordinary life there is no evidence that the drugs extend an individual's life expectancy. [8]

These are statistics coming from the latest reports. Some older reports had very telling statistics and I am sorry to report there has been little or no changes to the following statistics.
- John Cairns, M.D., Dept of Biostatistics, Harvard School of Public Health, notes that more than half of all cancer patients routinely receive chemotherapy drugs despite the fact that this form of treatment can be said to objectively help only a small percentage. [9]

[8] The Economist, October 20, 1990
[9] John Cairns, Scientific American, Nov., 1985

- John Bailar, III, M.D., stated, '"We are losing the war against cancer...this cancer data taken alone, provides no evidence that some 35 years of intense and growing efforts to improve the treatment of cancer has had much overall effect on the most fundamental measure of clinical outcome – death. Indeed, with respect to cancer as a whole, we are losing based upon the rise in age-adjusted mortality rates in the entire population."[10] (Obviously this has not changed by looking at the reports mentioned above regarding how far the U.S. has slipped in age-adjusted mortality)
- Robert Mendelsohn, M.D., Chairman of the Medical Licensing Committee for the State of Illinois stated, "There are plenty of ways that doctors produce disease. They will give you cancer-causing x-rays and say, 'you don't have to worry about it because it is low-level radiation.' Well, low-level radiation is a contradiction in terms. You might as well say a touch of pregnancy." [11]

We are constantly being told that we are winning the war on cancer, and that there have been improvements. This is just not correct. Earlier diagnosis has people statistically living longer because the cancer was discovered earlier. Many people have stopped smoking and taken other measures that have skewed the statistics regarding the success in cancer and many other problems.

We are being told by the emperors of medicine that everything is alright, and they are going to find the answers to our problems. It hasn't happened. And whether individuals have money or not does not mean they will be helped. The long list of affluent people including our most beloved movie stars that they could not pay their way to health. John Wayne, Michael Landon, as well as many wealthy corporate leaders, all succumbed to an aggressive attempt to address their cancer problems.

This book is not questioning what has not been done, rather it presents other technologies and therapies that offer greater possibilities. I am talking about approaches in the electromagnetic spectrum instead of merely chemical approaches. Speaking of those chemical and pharmaceutical approaches, the following chart appeared

---

[10] New England Journal of Medicine, May 1986
[11] Robert S. Mendelsohn, Confessions of a Medical Heretic, (McGraw-Hill, 1990)

in the January, 2009, issue of <u>Readers Digest</u>. The statistics were compiled by the Center for Responsive Politics. The figures are for federal lobbying through June, 2008. It is not too hard to figure why virtually no technique other then chemical is allowed by law.

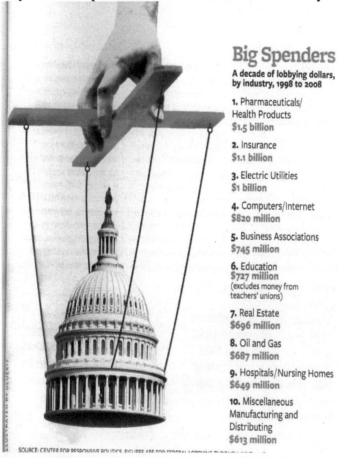

Big Spenders

A decade of lobbying dollars, by industry, 1998 to 2008

1. Pharmaceuticals/Health Products
$1.5 billion

2. Insurance
$1.1 billion

3. Electric Utilities
$1 billion

4. Computers/Internet
$820 million

5. Business Associations
$745 million

6. Education
$727 million
(excludes money from teachers' unions)

7. Real Estate
$696 million

8. Oil and Gas
$687 million

9. Hospitals/Nursing Homes
$649 million

10. Miscellaneous Manufacturing and Distributing
$613 million

SOURCE: CENTER FOR RESPONSIVE POLITICS. FIGURES ARE FOR FEDERAL LOBBYING THROUGH...

Shouldn't we take a closer look at what the emperor is wearing? I, for one, am tired of hearing the rhetoric that spending money researching approaches other than chemical will hurt the efforts of finding the answer through pharmaceutical research projects. I liken this to being placed on a new continent and being told to explore it but only in a northern direction. The northern direction, in this case, being a pharmaceutical answer.

So take a journey with me to discover some other proven approaches. The medical industry and the pharmaceutical industry

16

have virtually condoned only their chemical approach and stifled all competition to find additional answers.

Although this book discusses some forgotten technologies from the past, I am not endorsing or recommending any of them that might be currently in use somewhere in the world. Indeed some people have sought to capitalize on the name or popular opinion of something effective and proven in the past by creating their own unrelated therapy, giving it a similar name. Some dishonest people have built machines and used the name of some of the inventors whom we are talking about in this book. Then when these more modern machines do not work, the legal authorities place a ban or hold against an entire technology. They condemn the name of someone who lived several decades or more ago and most certainly did not build the current machine that is under scrutiny. My intention is to have these technologies reviewed, validated and proved effective again in our time. There is a game being played by some individuals within our "watchdog" medical groups where they like to say that something has been condemned or banned, and that shuts the door to further investigation of worthy technologies. This is disgusting and borders on criminal. To shut down an entire concept of electromagnetic effects upon the body because a person put out one machine that did not work or perhaps did work and was mislabeled, limits great discoveries that can be made.

Now it is up to the individual reader of this book to research very carefully before agreeing to any treatments utilizing the concepts that are listed within this book. I always caution people to ask manufacturers or therapists what research data is available on the specific machine that is to be tried. Further, when someone is taking the name of some researcher from the past and attaching it to his or her own technology, how closely is the device really aligned to the original concept?

If I have excited you enough to think about or question some of our current medical practice, then I have accomplished my goal. If I have not adequately stimulated you to understand the true potential of the techniques included in this book, then it is my fault due to a lack of writing skills and not the lack or true potential of the things mentioned herein.

YOU CAN MAKE IT HAPPEN!

Introduction
Part 2
Can We Afford The Tomato Effect Much Longer?

Robert W. Maver, F.S.A., M.A.A.A., was Vice-President and Director of Research, as well as the head actuary, for Mutual Benefit Life Insurance. The following is a personal research letter he wrote to me in 1989 that later appeared in "Discoveries In Medicine, Cost Containment Opportunities", from the Research Division of Mutual Benefit Life. The article follows in its entirety.

Maver

Can We Afford The Tomato Effect Much Longer?

"The premise is that there are innovative medical therapies existing today that offer solutions to some of our most pressing health problems and that at the same time offer significant reduction in health care costs. These therapies are largely being ignored or in some cases ridiculed."

"To most of us not involved in scientific research, this seems an odd notion at first. Surely, one would think discoveries and breakthroughs offering great promise in the treatment of disease would be at once communicated and embraced by the scientific/medical community. However, those who study the history of scientific progress conclude otherwise. Science frequently fails to demonstrate the dispassion we attribute to it."

"Historical citations of science resisting new ideas are too numerous to review in any depth, from Copernicus to Galileo to Darwin, Mendel, Ohm, Young, Harvey, Wegener, Semmelweis, Pasteur, Lister, Flemming...the list goes on and on. It is perhaps more instructive briefly to examine the reasons for resistance to innovation in medicine."

"Tomato Effect -- The tomato effect in medicine occurs when a highly efficacious therapy for a certain disease is ignored or rejected because it does not make sense in the light of accepted theories of disease mechanism and drug action. Doctors at the University of New Mexico School of Medicine introduced the tomato effect in the <u>Journal of the American Medical Association</u> (JAMA), May 11, 1984."

"Its name is derived from the history of the tomato in North America. By 1560 the tomato was becoming a staple of the continental European diet. However, it was shunned in America until the 1800s. Why? Because we believed it was poisonous. Everyone knew it. It was obvious. Tomatoes belong to the nightshade family. The leaves and fruit of several plants in this family can cause death if ingested. The fact that Europeans were eating tomatoes without harm was not relevant. It simply did not make sense to eat poisonous food."

"**Peer Review** -- The peer review process has probably done more to discourage innovative research than any other factor that I have observed. The March 9, 1990, issue of JAMA was devoted entirely to the topic of peer review. One article in particular, by Horrobin (himself editor of a peer reviewed medical journal) cited 18 examples of peer review attempting to suppress medical innovation. The article observed: '…some of the most distinguished of scientists may display sophisticated behavior that can only be described as pathological. Editors must be conscious that, despite public protestations to the contrary, many scientists-reviewers are against innovation unless it is their innovation. Innovation from others may be a threat because it diminishes the importance of the scientist's own work.'"

"Peer review in the grant giving process is so restrictive that most innovative scientists know they would never receive funding if they actually said what they were going to do. Scientists therefore have to tell lies in their grant applications. Such views have been explicitly stated by at least two Nobel Laureates."

"This article contends that medicine has lost sight of the basic purpose of peer review, asserting 'the true aim of peer review in biomedical science must be to improve the quality of patient care.'"

"**Wrong Economics** -- When a new therapy comes along that is cheaper, safer and more effective, it is seen as a competitive threat by those engaged in the therapy it will displace. Those who stand to be most economically disadvantaged naturally endeavor to block its acceptance."

"**International Barriers** -- A combination of communication problems (language barriers) and national chauvinism (if it wasn't discovered here it can't be of much value) keep some innovative practices developed in Europe and Asia from reaching the United States."

"**Cumbersome Bureaucracy** -- It has been estimated that the FDA approval process takes an average of 12 years and costs $231 million. This presents unique difficulties for independent researchers and for therapies that do not lend themselves to patentability."

"It is my observation that there is a role for the insurance industry in advocating evaluation of innovative medical therapies. Actuaries should be almost immune to the tomato effect. We are focused almost exclusively on statistical results as opposed to theory. Since the insurance industry pays most of the bills, it should have great economic motivation to see safe, effective and inexpensive therapies extensively evaluated and widely disseminated. Consideration of an industry-wide fund for innovative research could deal with the problem of peer review. The insurance industry is a sleeping economic giant. When it awakens to the cost containment possibilities available through innovative therapies, we will see enormous changes in the practice of medicine."

I could not agree more with Bob's comments regarding what is taking place in our medical system. It is an unfortunate aspect of life that greed, ego and power seem to come into play in most aspects of our lives. I believe that we can live with these things when we are talking about various products, goods and services, but not when it comes to our life. This is when and where the line must be drawn in acceptance of these factors.

In this era of big bucks and the use of our public relation firms to create big reputations that will lead to greater rewards, we have lost some, if not most, of our spiritual and truthful principles. When I read in our newspapers that some research trial was stopped because there were too many deaths, I weep. We will march against animal trials but we accept these human casualties in the name of research.

Nikola Tesla, one of the greatest scientists of any age, remarked that his inventions followed the laws of nature. He was not fighting nature and trying to force or manipulate nature to create something. He observed nature and arrived at the greatest inventions of our time. This should be our creed in most areas of our lives. It is not about force, manipulation and pushing, but about harmony, assisting and supporting.

Introduction
Part 3
How Do We Treat The Messengers - Past, Present and Future?

Ignaz Philipp Semmelweis, who was called the savior of mothers, was born in Buda, Hungary, on July 1, 1818. He was educated at the universities of Pest and Vienna. In 1844, he received his doctor's degree and ended up with an appointment as the assistant at the obstetric clinic in Vienna.

From very early in his career he became interested in puerperal infection, better known as childbed fever. This was a situation of almost epidemic proportion in the maternity hospitals throughout Europe but was only rarely found in a home delivery. Although the majority of the women during the period delivered at home, some women had to go to hospitals because of poverty, or their children were illegitimate or there were problems with deliveries.

Dr.Ignaz Semmelweis.
By Jenő Doby,

The mortality rates were as high as 30% in the hospitals. Many people believed that the problem was due to overcrowding and poor ventilation that were prevalent in the hospitals. Semmelweis would not accept this situation without attempting to find an answer, despite strong objections by his superior Professor Klein. Klein had reconciled himself to the fact that this situation was always going to exist.

Semmelweis began his investigation by noting that within his own hospital women in the First Division had a death rate from puerperal infection two to three times greater than women who were delivering in the Second Division. Although the divisions appeared to be virtually identical, he did note that medical students received training in the First Division and midwifes received training in the Second Division.

Although he had been aware of this difference, he did not find his answer until a friend and colleague of his died. The friend died of a wound infection that occurred during the examination of a woman who had died of puerperal infection. The autopsy of this man was strikingly similar to that of the woman whom his friend had autopsied. He eventually concluded that students were coming from the dissecting room and going directly to the maternity ward. He understood that infectious materials were being transmitted from diseased corpses to healthy mothers.

Semmelweis demanded that all doctors and students wash their hands before examining any patient. He suggested a solution of chlorinated lime. The results of his new procedure were staggering. The mortality in the First Division dropped from 18% to approximately 2%, and during two months in 1848, no women died in childbirth in his division.

Despite his wonderful success, his supervisor Professor Klein would not accept the findings and openly ridiculed him. The students and doctors had mixed emotions but tended to side with the supervisor and the old thinking. In 1849 Semmelweis was dropped from his post in the hospital. He moved to the city of Pest and worked at the St. Rochus Hospital.

There he implemented his procedures. Promptly the mortality rate declined. In his obstetrics department the mortality rate was less than one percent. In Prague and Vienna during the same period, the rate was approximately 15%.

His ideas were accepted in Hungary, and the government sent an edict to the district authorities ordering that they use the prophylactic methods of Semmelweis. Other areas were still openly hostile. In Vienna, the medical magazines were writing about the ridiculous nonsense of Semmelweis.

In 1861, Semmelweis published his classical work on the subject. He received adverse reactions, and he was not received well by any of the professors of medicine or the majority of the natural scientists of his time, notably the famous pathologist Rudolf Virchow who rejected his theory and procedures.

In 1865, Semmelweis suffered a breakdown and ended up in a mental hospital where he was placed in a straightjacket and eventually died. Ironically, his death was caused by an infection of a wound on his right hand, apparently as a result of an operation he had performed. He died of the same condition against which he had struggled during his career.

Eventually his doctrine was accepted by medical science. Joseph Lister, the father of modern antisepsis stated, "I think with the greatest admiration of him [Semmelweis] and his achievement, and it fills me with joy that at last he is given the respect due to him."

The Semmelweis story is very sad, yet an indicative reflection on how the world has treated those individuals who dared to challenge the orthodoxy of their day. Often, it doesn't matter whether someone is presenting a truth; success or failure is predicated upon how many "boats are rocked." In 1592, during the Italian inquisition, Girodano Bruno was imprisoned for eight years and then burned alive at the stake. During his life, Bruno related his philosophy of an infinite universe where the earth was a star and moved around the heavens like other planets. He worked with the power of love and imagination in healing. Bruno discussed how the ancients had used mythology to normalize psychological problems of the human mind.

It would be nice to believe that such harsh treatment to the messenger of new ideas is a thing of the past. But sadly it is not!

I have compiled and documented numerous examples of medical doctors and scientists, from even the last 30 years, who have been subjected to harsh repercussions for coming out with some new therapies or techniques that were not parts of accepted medical norms.

Paracelsus, the great philosopher and medical thinker from the early 1500s, stated that in Europe it was virtually impossible to come out with anything that was new or innovative in medicine. The reason was that it would upset the theory of some professor of medicine in one of the universities, and it would not be allowed to challenge that authority figure. This is why Paracelsus was  able to learn and use his creative thought in Constantinople. It was there that Paracelsus said the people were receptive and liked to learn and discover for the sake for truth. There were not as many *sacred cows* to overcome as there was where he practiced in Europe.

Manly Hall, a great philosopher of our modern times, has remarked that in ancient times people were considered great because of what they knew and what they taught. In today's world we tend to consider people's words of great authority mostly because of their credentials and the number of titles and degrees they hold. This is not far from the truth. I would trust a Semmelweis who had practical experience above someone revered simply for the initials following his or her name.

Perhaps this is why Paracelsus once stated, "the way to learn nature is to walk her leaves." It is by practical experience and life's lessons, in addition to academics, that we can learn the highest truths.

When I hear the answer to solving a health issue is right around the corner and all we need is to raise money to find the pharmaceutical cure, my answer is: The answer is not right around the corner; it is already sitting right here before us, and it may not require pharmaceuticals at all. Certainly, some pharmaceuticals are lifesavers. My concern, of course, is when medicines are not only not effective, but also dangerous. There would appear to be many non-pharmaceutical approaches that are indeed effective with no harmful side effects.

I do not wish to attack the medical practitioner who vigilantly seeks to heal, but rather the education process that teaches our doctors to look for the right chemical to prescribe rather than following the dictates of nature. It has been attributed to Hippocrates that he made the statement, "First do no harm." Whether he made the statement or not, it seems a wiser course to follow than our approach of accepting side effects from our therapies.

Chapter 1
Preparing For My Journey

During my years of working in the health field, I have met and counseled many people who shared sad and heartbreaking stories regarding their upbringing and their interaction with family members. Often, after successful sessions, a person might turn to me asking if I had a harrowing or shocking time during my youth. I believe people assume this because I attempt to get into a sympathetic rapport with those whom I meet.

It appears that I am one of the more fortunate people because I did not have any dire experiences while I was growing up. I grew up in a home filled with music, love, beauty and laughter. There was always great stability at home, at least from my perception at the time, and my brother and I never wanted for anything. My mother and father, who would eventually be married for 57 years, were always hugging and kissing each other, and my mother expressed the same physical contact to Barry and me.

Although it seems like a blur at this moment, my father told me that early in life I was quite mechanical. That would really be a joke to my friends who see me as very much the opposite in my skills. My father was proud of telling the story how, for an elementary school project, I constructed a unique wood house and then showed the effects of what happens when a fire takes place. I built two models that my father claimed were quite good. Seeing that my father was an engineer and master mechanic, this was a great compliment.

I seemed to have lost this ability when I was twelve years old. Today, I am at the mercy of every mechanic, electrician, plumber and builder. Items that I build with my hands do not seem to last long and are a bit shaky. You would not want to sit in a chair that I built from those kits that claim easy assembly.

Several years ago, I made a last ditch stand to regain some credibility with my family. I purchased a gumball machine for my brother and decided that I would put the parts together before wrapping it up for Christmas. I should have known there might be a problem when I discovered that there were two screws left over after it appeared to be completed. It was disheartening for me when my brother opened the wrapping paper and let me know that the gumballs would never roll out of the contraption that I had assembled for him. I had screwed the actual gum container slide upside down! I salvaged some dignity with the thought that I had at least tried.

When I was about nine years old, I remember asking my parents what religion we were. They swiftly replied, "What religion would you like to choose?" My profound, intelligent and well-thought answer: "I do not know." With that response, my parents told me that I was free to choose any religion if I could explain to them what the religion was and why I wanted to follow it. When I asked them how I would know which religion was best for me they responded that it would come through my learning and the experiences I had.

To get me started, they brought me several different books on religion and philosophy. My parents told me to read, be aware of feelings that I had and question whatever I might find. They again mentioned that whatever I decided to choose would be acceptable to them. They encouraged me to attend any religious service that a neighbor or friend might ask me to attend. In this way, I would experience the religion first hand.

As we continued to have deeper discussions, my parents expressed their belief that following the tenants of the Ten Commandments would always be good guidance.

When I was approximately 15 years of age, I asked my parents what our political party was. Their answer, "What political party do you believe is the best?" I told them that I did not know. They said that I should read what each candidate was saying and make my choice based on what I believed would be the most helpful for all people. They proceeded to supply me with more information than I had intended to read, but it allowed me to have a better understanding of the process of choice.

These early experiences with religion and politics formed an important aspect of my character. I was not limited to one choice but was presented the opportunity to make decisions based on as much evidence as I could ascertain.

What a blessing to have parents who were quite open to all things and were not dogmatic. My beautiful parents provided me with constant love and encouragement to find my own truth and taught me not be afraid to experience new things.

I mentioned in the preface that I had a interesting sports background. My sports experience taught me that a person can expend tremendous energy, desire and will, but that it might not lead to the expected outcome. During my college career, I participated in several indoor track and field competitions that taught me the valuable lesson of perseverance. Although my best events were 100 meters and 200 meters, during the winter indoor season I would often compete in 50 or 60 meter races. These shorter distances were highly competitive and the results were always determined by a photo finish to determine the winner. The finish line judges would need a photo to determine whose chin, nose, chest or arm crossed the line first. A season of preparation would come down to winning or losing by a margin that was quite small. Whatever the outcome, I would have to prepare for the next weekend of competition.

Today we read of trash-talking athletes who seem to derive pleasure out of belittling their competition and making fancy body gyrations after they win. During my competitive days, I preferred to measure myself against my own results. If I ran a better time in my event than I had run in a previous competition then I considered myself a winner no matter what my competitors did. This understanding helped me a great deal when I broke my university school record but another athlete in the race tied the world record. It was the largest distance I had ever been behind another athlete in my event, and I watched this athlete's backside for 100 yards. My coach had a great attitude and was excited that I had broken our university record. Everything is a matter of perspective, and I was still a winner.

As I mentioned in the preface, I sustained an injury during competition that peeked my interest in the health field. Although this was not my college major, it would become my future passion.

During high school, I was a math/science major, and I graduated from the university with a degree in business. During my mid-teens, I became interested in the stock market and actually spent time composing stock charts for several brokers. This was in the mid-1960s, and my parents had to sign for me when I made any personal investments.

In those years, I assumed I would become a stockbroker. When I graduated from college, I took some examinations with a large brokerage house in the Los Angeles area. I was offered a job and was told that within the next three weeks I would be sent to New York for training. That never happened because one week later three brokerage houses went out of business, and many experienced brokers with clients were now available. I received a call from the brokerage company telling me that I was a nice young man but they had to go with more experienced people. That ended my career in the stock market.

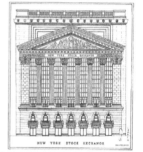

Because it did not appear that I was going to be a stockbroker, I spent the next two years doing some business consulting along with learning more about therapies in the health field.

My aim in writing this book is not about <u>how</u> I discovered the materials I have included in the book. The most important aspect is <u>what</u> was discovered and how you could benefit.

Trusting my own inner awareness and vision is what has led me to investigate, seek and make the discoveries that I have written about in this book. My search has taken me all around the world, and I am very grateful for the life I live. When I envision myself sitting in an office watching the stock market go through its gyrations in an unending swing, where people make and lose fortunes in just minutes, I can't help believe that I was blessed with everything that has happened to me.

I have personally benefited from many of the materials that are included in this book, so it is a bit of a misnomer when I say, *"and nothing happened."* This comment is relative. Just the knowledge and use of these materials have personally benefited my life. My goal is to help these materials benefit your life.

ONLY YOU CAN MAKE IT HAPPEN!

The only limitation is what we conceive it to be.

Chapter 2
Surpassing All Limitations

For several years, I had been looking to find an out-of-print book that had good information on a variety of alternative healing techniques. In 1983, we did not have the wonderful internet locations and book dealers we have today that makes it easier to find these types of books. I was very fortunate growing up to have access to a fantastic bookstore called the Bodhi Tree. The Bodhi Tree was a treasure for anyone who liked books dealing with health, philosophy and spiritual topics.

One night in 1983, I had a dream that I was in the Bodhi Tree bookstore and came to a particular section of books. I looked at the books on the shelf but did not find any that interested me. However, when I reached behind the row of books, there was a book lying behind the stack.

I was excited and could not wait until the bookstore opened the following morning. I went to the section that I saw in the dream and when I reached behind the stack, there was a book lying on its side. *New Light on Therapeutic Energies*, by Mark Gallert, was the exact book that I had been looking for!

I had been doing some research regarding electromagnetic energies and their relationship to health, and this book had several topics that were of interest.

One chapter dealt with a topic that I had never thought about or had any previous interest in. It was information regarding a powerful microscope developed in the 1920s. My first inclination would have been to bypass the chapter for some other topics that appeared more interesting. However, I did decide to read the chapter.

The author of the book had only written about six pages on what he called the Rife Microscope, but those six pages absolutely overwhelmed my senses. The book stated that the Rife microscope, built in the late 1920s and early 1930s, was ten times more powerful than the light-source microscopes that were still in use by the 1980s. The book stated that through this microscope the researchers had discovered something that would change the entire course of medical theory and therapy in the world. Apparently, the microscope disappeared because the paradigm change would have altered the use of pharmaceuticals in medicine.

There was more information, but at the time it was somewhat over my head. The whole subject had come out of nowhere, and I did not feel like concentrating enough to get the full impact. I was intrigued because one line in the book stated that the microscope defied all the known laws of optics and far surpassed the limitations that science had set in microscopy.

As I digested most of the other sections of my new book, I decided to spend more time to understand what the importance of this microscope could be. I decided to check with a medical doctor who was a friend of mine. Dr. Harry Lusk was in his early eighties and had been the head of the gynecology section of one of the largest hospitals in the Los Angeles area. He still had a private practice, and if anyone in my circle of acquaintances would know about this topic, perhaps it would be Harry.

I called Harry and asked if we could meet for lunch. We set a day and time that I eagerly awaited. During our lunch, I brought up the Rife Microscope and asked Harry if he had ever heard of it. I was surprised because he responded almost instantly that he had heard about it. Dr. Lusk told me that when he was a young student in medical school he had heard rumors of a super microscope built in California, and that it was different than any other microscope in existence. Harry went on to explain that there are big differences in microscopes. Medical research uses light-source microscopes, and they have a magnification of about 1800 diameters. Although electron microscopes might have enlargement up to one or two million magnification, researchers are not able to view live specimens. This is because a specimen placed in a vacuum and bombarded by electrons would not remain alive to view. Therefore, only the light-source microscopes that we have all seen in school are the tools used to do medical research.

Harry went on to tell me that there was a limitation according to science regarding the use of light and, therefore, all the light-source microscopes were limited to this magnification. Apparently, the Rife Microscope surpassed this limitation.

I wanted to know more, but Harry did not know much more. He told me that about ten years earlier, in 1975, he had attended a talk given by a man who knew all about the Rife Microscope. This fellow's name was John Crane. Harry remembered that John had befriended Dr. Rife before he died and actually had the Rife Microscope in his possession. I asked Harry why he had not followed through to learn more about the story of the microscope and what really happened during those early years. Harry told me that he had tried several times a few months after the lecture, but John Crane never answered his telephone.

I begged Harry to give me some type of contact information. Harry told me that it had been a good eight or nine years, and he had no idea where the contact information was located. I asked him to please look to see what he might find. Harry reminded me that he had recently moved, and he did not believe he could find anything in his home.

The next morning I was very surprised to hear Harry's voice, especially when it was quivering as he spoke. He told me that when he had gotten out of bed in the morning he had stepped on a paper. The paper had John Crane's telephone number and a notation about the Rife Microscope. I joked with Harry that it was probably there all the time, and he assured me that it was definitely not there before we had spoken.

He just could not figure out why this happened as it did, and he was looking for some deeper explanation when I stopped him and asked if he would give me the telephone number. He again reminded me that nobody answered the telephone. He made me promise that if I did make contact and was going to see the microscope that he would come with me. He told me that John Crane had lived in San Diego, and we could make that trip together in just a few hours. I promised, and he gave me the number.

I made the telephone call and a man immediately answered the phone. I asked if John Crane was there but, the response that I received was a list of questions regarding why I was calling this number. For thirty minutes, question after question bombarded me regarding whether I was working for the FDA, the AMA, the FBI or any other governmental organization.

I became a bit apprehensive of what I was doing pursuing this microscope because this was paranoia at a level that I had never experienced. Finally, this intense grilling concluded when I mentioned I had co-founded a worldwide health organization to protect these type of discoveries.

Indeed, I had reached John Crane, and he invited me to come to San Diego to meet him. I told him I would be coming with my friend Dr. Lusk, and we would meet in three days. When I told Harry, he was ecstatic and could not wait for our journey.

Early on the morning we were to leave for San Diego, Harry called and said that one of his patients was having her baby and he was at the hospital. It was a happy event for the mother but not for Harry and me. I called John and told him we could not make it that day and please do not think I was a flake. Could he give us another meeting date? John was very nice and said that he could meet on the upcoming Saturday.

I do not know why I felt better knowing it was a weekend, but Harry called on that morning with the news of another imminent baby delivery. Why did I assume babies do not pop in on weekends! Even though I promised Harry we would visit together, I left him and made my way to San Diego.

When I arrived at the house, it appeared as if no one was living there. Approaching the front door, all the windows looked tightly closed, and the grounds were not in good shape. A few moments after I knocked, the door opened just a crack. Steven Ross, the voice asked? It was John Crane, and he let me inside.

Although it was late morning, the window shades were tightly closed and there was no outside light coming into the room. There were newspapers all over the main room, and the level of dust was almost overwhelming. John asked me to sit on a sofa, and plumes of dust rose into the air. Now I was glancing at an empty glass on the table in front of me when John asked if I wanted some water. The glass looked like some chemistry experiment; I told him that I would pass on any refreshment.

John related some of the story of Royal R. Rife and the super microscope he had designed. John was familiar with the story of Rife because, in the 1950s, he had met Royal Rife; and when Rife passed away, he left John all of the materials, including Rife's super-microscope. At that moment I wanted to see the microscope so bad, I was starting to

Universal Microscope

squirm. John picked up my actions or thoughts and said, "Follow me." As we entered the room he took me to, on a table in the corner was the Rife Universal Microscope.

The microscope was sitting inside a clear glass case. It was so impressive to look at that it almost really didn't matter what it did. It looked like something you would see in a Star Wars movie. It felt like a movie where someone finds the Holy Grail, or perhaps I was an archeologist who unearthed some hidden mystery. I could feel history attached to this device.

During the next several months, I returned to San Diego, and John told me more facts regarding the story of Royal Rife and his Universal Microscope. Interestingly, Harry never made it to San Diego to see the microscope, but I did bring LaVerne, the co-founder of World Research Foundation on several trips. After many meetings, John called me and told me to bring a large car or station wagon on the next visit. On this historic visit to San Diego on July 18, 1984, John entrusted me with the Rife Universal Microscope.

That day I signed a document acknowledging that I was a caretaker of the microscope. In addition, I was given some originals and copies of all of the letters, papers, documents, pictures, laboratory notes and other items of Rife's that were in John Crane's possession. I will never forget driving on the freeway from San Diego to Los Angeles looking in my rear view mirror and seeing the Rife Microscope in the back of my station wagon.

Today, the internet is loaded with information regarding the Rife Microscope, but virtually none of the authors of these writings has ever physically seen the Rife Microscope. People seem to pass along stories and information that they copied from another internet site. When I obtained this information, the internet did not exist in its present form, so the following facts I am sharing are from Rife's original materials I obtained from John Crane.

This is not a scientific book nor am I going into detail on some of the points. What is most important is the overall concept of what would change in our world if something had happened with this microscope.

Royal Rife worked in San Diego as a chauffer for Mr. Timken, a manufacturer of ball bearings for carriages. Timken was very impressed with the creative mind of Rife, and in the early 1920s, built a research laboratory for Rife valued at one million dollars.

Rife was fascinated with the idea of microscopes and magnification and did not believe that the commonly accepted scientific laws dealing with light beams and magnification were correct. Rife proved his theories when he invented several different microscopes; each surpassed the theoretical limitations of microscopy that even today is still accepted as a dogmatic fact of science.

Rife's Universal Microscope was able to achieve magnification of 50,000 diameters, where all other microscopes could obtain only 1,800 diameters. When you compare these different magnifications, can you imagine what a different viewpoint Rife would have versus what other researchers were seeing?

Today, virtually every research center on the planet earth is still using the same low magnification microscopes. The question is why they are using the same devices used in the past and not this super microscope. The reason is what Rife and his associates discovered through their use of this powerful microscope. Interestingly enough, their greatest discovery was not made because of the increased magnification that Rife achieved but from another principle that he was using.

I have mentioned that Rife had several associates who were working with him, and perhaps it is a good time to mention these other medical specialists. **You will find copies of many of the following letters and newspaper articles in the appendix of this book**.

On November 9, 1931, Dr. Milbank Johnson of Pasadena, California, wrote a letter to Rife acknowledging that his visit to Rife's laboratory was the "most interesting afternoon of my life." Dr. Johnson was the head of the American Medical Association's local center in Pasadena. Dr. Johnson was a highly respected medical doctor and would later become a founding member of the Automobile Association of America.

Dr. Johnson's contacts extended throughout the United States and included doctors from famous medical institutes and scientific research centers. Dr. Johnson immediately contacted his close friend Dr. Arthur Kendall of Northwestern University who had written several medical textbooks on bacteriology that were used in most medical schools.

After Dr. Kendall's visit to Rife, Kendall was so overwhelmed that he contacted his close friend Dr. E.C. Rosenow of the Mayo Clinic. Dr. Rosenow would write in a letter to Rife on July 11, 1932: "After

MILBANK JOHNSON, M. D.
PACIFIC MUTUAL LIFE BLDG.
LOS ANGELES, CALIFORNIA

November 9, 1931

My dear Mr. Rife:

In the name of the other three gentlemen and myself I want to thank you for your most courteous reception and for giving us an opportunity to have a glance of your wonderful microscope. I want to say to you that we all spent one of the most instructive and interesting afternoons of our lives in your laboratory.

Upon returning to San Diego that evening I wired to Dr. Arthur I. Kendall of Chicago and gave him a brief description of what we had seen and our opinion of it, and upon my return to Pasadena this morning I received the following telegram from Dr. Kendall - "Expect to start for California Saturday night. Letter follows".

If he comes straight through, which I think he will, he will arrive in Pasadena on Tuesday, November 17 so be sure and have your microscope in perfect condition for the Big Chief when he arrives. I will bring him down to San Diego in my car at which time you and Dr. Kendall can make such arrangements as you desire.

Thanking you again for your courtesy, I am

Yours very sincerely,

Milbank Johnson

600 BURLEIGH DRIVE
SAN RAFAEL HEIGHTS
PASADENA

Mr. Roy Rife
2500 Chatsworth Bldg.
San Diego, Calif.

THE UNIVERSITY OF MINNESOTA

THE MAYO FOUNDATION
FOR MEDICAL EDUCATION AND RESEARCH
ROCHESTER, MINNESOTA, U.S.A.

July 11, 1932

Dr. R. R. Rife
712 Electric Building
San Diego, California

Dear Dr. Rife:

After seeing what your wonderful microscope will do, and after pondering over the significance of what you revealed with its use during those three strenuous and memorable days spent in Dr. Kendall's laboratory, I just must again express the hope that you will take the necessary time to describe how you obtain what physicists consider the impossible as regards magnification, and submit your paper to a suitable Journal for publication. Dr. Kendall, I'm sure, could give you the best of advice as to where it should be published, and if you would like to have me, I would be glad to help you also.

As I visualize the matter, your ingenious method of illumination with the intense monochromatic beam of light is of even greater importance than the enormously high magnification obtained with your present instrument. I hope that you will be successful promptly in your attempt to develop a similar method of illumination which will make it possible to see these "filtrable" forms with the oil immersion magnification.

The outstanding result of the observations made in triplicate during those three days, it seems to me, is the fact that the "filtrable" forms are approximately as large as the visible, stainable forms. In other words, it appears that filter-passing forms are perhaps not extremely small but are more plastic or perhaps immature forms, and by virtue of this state rather than because small, are drawn through the pores of the filter. Observations made with filtrates of cultures of the streptococcus from poliomyelitis and of the cultures containing the contaminating diplococcus, and those with filtrates of poliomyelitic and herpes encephalitic virus from which I commonly isolate streptococci, have added much, it seems to me, in the way of the correctness of your and Dr. Kendall's observations on the filter-passing forms of B. typhosus.

A new point of view has been opened up and I wish so much that you could be here to parallel our observations during our study of poliomyelitis and our attempts to develop the virus from the streptococcus.

I was most favorably impressed with the results obtained with seeing what your wonderful microscope will do, and after pondering the significance of what you revealed with its use during those three strenuous and memorable days spent with you and Dr. Kendall...I must again express the hope that you will take the necessary time to describe how you obtain what physicists consider the impossible as regards to magnification...and submit your paper to a suitable journal for publication."

Along this same theme of obtaining what others felt was impossible, Dr. Lewellys Barker of Johns Hopkins would tell the San Diego Sun newspaper on January 7, 1932, that he was very impressed with Rife's instrument and that the scientists who were claiming the device is violating all the known laws of optics would have to get some new laws.

On November 27, 1931, a representative of the Spencer Lens Company wrote a letter to Rife stating that he had heard from a group of his scientific friends who attended a demonstration of the Rife Universal Microscope at California Institute of Technology (Cal-Tech). In his words, it left all of the attendees, "agog."

Royal Rife in his laboratory

In the most simplistic of terms, this is what Rife and his associates discovered. All the little microbes, bacteria, and germs that they viewed through the microscope gave off different colors of light. The same type of germs, bacteria or other microbes gave off the same color every time they were viewed throughout their lives. These colors existed in an ultra-violet spectrum, but because Rife superimposed another beam of ultra-violet light, they would produce the colors that were seen through the microscope. Therefore, it wasn't just due to the increased magnification but because of the principles that Rife applied that this effect occurred.

This is important because every color that we see is merely a rate of vibration that produces that color. Different rates of vibration produce different colors. As vibrations increase or decrease, the colors that you perceive change in their appearance to your eyes.

Further to this discovery, as the researchers were investigating the causes of different diseases, illness and health conditions, they would see various organisms of different colors. The same medical conditions always had the same specific colored organisms present with the specific medical condition being examined.

Using the increased magnification that could only be seen under Rife's Universal Microscope, the researchers were able to see viruses and could make the following conclusions:
The virus of Vacillus Typhosus was always a turquoise-blue, the Cacillus Coli was always a mahogany-color, the Mycobacterium li prae was always a ruby shade, the filter-passing form of tuberculosis was always an emerald green and the virus of cancer was always a purple-red.

Royal R. Rife and Universal Microscope

MANUFACTURERS OF
HIGH GRADE
OPTICAL INSTRUMENTS

Spencer Lens Company
Factory, Buffalo, N. Y., U. S. A.

SOUTHERN CALIFORNIA BRANCH:
605-6 BEAUX ARTS BUILDING
1708 WEST EIGHTH STREET
LOS ANGELES, CALIFORNIA

PHONE: DUNKIRK 7575

LOS ANGELES, CALIFORNIA.

November 27, 1931

Mr. Roy R. Rife
712 Electric Bldg.
San Diego, Calif.

Dear Mr. Rife:

Just a short personal line to tell you that you have
made a very favorable impression on the scientific
people in and around Los Angeles. We recently heard
about a demonstration that you made at the California
Institute of Technology and many of my friends connected with the educational institutions have spoken
to me about the demonstration. It certainly has them
all "agog."

I also wish to extend to you my sincere thanks for
the very kind interview and time that you gave to a
very dear friend of mine, namely, Dr. Charles Chamberlain of the University of Chicago. Dr. Chamberlain
is well liked and loved by all who know him and you
have made an old man very, very happy.

With kindest personal regards, and assuring you of
my best wishes for your success, I am,

Yours sincerely,

Lyle D. Potter

LDP-GEA

Using the Rife microscope, the researchers would find a typhoid organism in the blood of a suspected typhoid patient four - five days before the Widal test would be positive for the problem.

The doctors proved through their experimentation that the various colored microorganisms were always present when there was a health or medical problem. Rife eventually had the realization of how he could destroy, kill or render these organisms harmless. His thought process might have been along the lines of what happens when a singer sings and hits a note that is the same as the resonant or inner frequency of a glass. It will shatter. What happens if a group of soldiers marching across a bridge happen to produce a frequency similar to the resonant frequency of the elements that make up the bridge? The bridge will collapse. What happens when two automobiles strike each other head on at exactly the same speed? They will cancel each other out. These examples all have to do with the law of vibratory resonance.

Rife and his associates contacted Dr. Le DeForest who was a Yale University Ph.D. and was a pioneer in the development of radio communication. Dr. DeForest's most important invention was a type of vacuum tube that was called the audion, which today is known as the triode. This tube, invented in 1906, revolutionized the entire field of electronics. De Forest's knowledge of electronics coupled with the discoveries of Rife and his associates could allow the creation of a new therapy device.

Using a frequency-generating device designed by Dr. DeForest and his unique microscope, Rife would eventually discover the specific frequencies that would destroy individual microorganisms. The ultrasound frequencies would literally shake the microbe until it was destroyed. Rife would discover the single frequencies of microbes that were associated with more than 50 different diseases and illnesses.

Some twelve years later, a report in the Smithsonian would state, "Under the Universal Microscope disease, organisms such as those of tuberculosis, cancer, sarcoma, streptococcus, typhoid, staphylococcus, leprosy, hoof and mouth disease and others may be observed to succumb when exposed to certain lethal frequencies coordinated with the particular frequencies peculiar to each individual organism. It should be emphasized that invariably the same organisms refract the same colors when stained by means of this mono-chromatic beam of illumination." [1]

Rife discovered that healthy cells in the body were not in any way affected by the frequencies that he used to destroy the harmful microorganisms. The cell wall of a healthy cell has no problem withstanding the frequency that destroys the harmful invading microbes.

After more than 10,000 experiments with test animals, the researchers were completely convinced of the efficacy of their approach and were determined to investigate the effects on patients. To accomplish this, Dr. Milbank Johnson contacted his close friend, Dr. Rufus Von Klein Smid, who was the president of the University of Southern California. They established a Special Medical Research Committee at the University of Southern California. Letters from the period between 1933 and 1937 show the following members of the committee:

Dr. Milbank Johnson, Chairman of the Board of USC.

Dr. Rufus B. Von Klein Smid, President of USC

Dr. George Dock, Prof. of Medical Department Tulane University

Dr. Charles Fischer, Children's Hospital of New York

Dr. Wayland Morrison, Chief Surgeon, Sante Fe Railway Hospital

Dr. Karl Meyer, George Hooper Foundation, Univ. of California

Dr. E. L. Walker, University of California

Dr. Alvin G. Foord, Pres., American Association of Pathologists.

Dr. E.C. Rosenow, Head of Research, Mayo Clinic

Dr. Arthur Kendall, Northwestern Univ., First Chair of Chemistry at Harvard Univ.

Dr. O. C. Grunner, Archibald Cancer Research, McGill Univ.

---

[1] R. E. Seidel, M. Elizabeth Winter, "The New Microscopes", *Annual Report of the Smithsonian Institute* for 1944, p. 207

MILBANK JOHNSON, M. D.
PACIFIC MUTUAL LIFE BLDG.
LOS ANGELES, CALIFORNIA

October 8, 1935

My dear Dr. Rife,

We are about ready to begin our clinical work with the new Rife Ray Machine which seems to be a great success. It has much greater power and penetration than the original which we used last summer.

There are many improvements in this machine which are possible through the great improvements made in radio technique. There is not a moving part, for example, in our new machine and hence we expect it to have a much longer life with harder usage.

We believe it wise to protect the members of the Committee and the physicians from suits for damages. Your Chairman, therefore, has had prepared by experienced lawyers two forms of release which I am submitting to you for your suggestions or approval. Kindly read them over very carefully. Consult any attorney you please if you so desire, and return them to me as promptly as possible as we are about ready to start.

We have tested the machine out very thoroughly both on animals and on cultures, and so far as we can see, it leaves nothing to be desired.

Hoping that you will examine and return the releases to me with your comments as quickly as possible, I am

Yours very sincerely,

Milbank Johnson, Chairman

Special Medical Research Committee of
the University of Southern California

Dr. Royal Raymond Rife
2500 Chatsworth Boulevard
San Diego, California

43

MILBANK JOHNSON, M.D.
PACIFIC MUTUAL LIFE BLDG.
LOS ANGELES, CALIFORNIA

December 19, 1935

My dear Dr. Rife,

      A meeting of the Special Medical Research Committee of the University of Southern California will be held Thursday, December 26 at 12:15 P.M. in Room 2 of the California Club.

      As Dr. George Dock, a member of our Committee, is leaving on January 2 for a trip around the World and will not return for several months, I am anxious to have this meeting before he leaves as there are many things of importance to be considered. We have much to report and are very anxious to receive your advice on some questions of vital importance to the work.

      I trust you will make a special effort to attend. I have tried to trouble the members of the Committee as little as possible with meetings, but it becomes absolutely necessary now that we should meet and decide some vital points.

      You might call Dr. Burger and see if you can't arrange to come up together as you did last time. Also, it will keep him from forgetting it and insure his being here if you bring him.

      Please let me hear from you as to whether or not you can be present at this meeting.

      Wishing you and Mrs. Rife a Merry Christmas and a Happy New Year, I am

Yours very sincerely,

Milbank Johnson, Chairman
Special Medical Research Committee
University of Southern California

Dr. R. R. Rife
2500 Chatsworth Blvd.
San Diego, California

600 BURLEIGH DRIVE
SAN RAFAEL HEIGHTS
PASADENA

Testing and experimentation began with the specific frequency rates called MOR that were discovered by the research team. MOR stood for the Mortal Oscillatory Rates that Rife found using test animals and blood samples from patients. I have a video that was taken in the 1930s showing Rife working with the test animals in his laboratory. When I met John Crane this movie was on an old reel-to-reel tape. I was the person who converted this to a VCR for John. I know of many individuals who now have this wonderful video showing the extensive amount of laboratory equipment that Rife had available for his research. One of Rife's experiments is so dramatic that it needs to be shared.

Rife received a tumor that was taken from a woman's breast. The tumor was cancerous. Rife cut the tumor in half and used one half to infect a series of test animals with the cancer. Every one of the test animals developed cancer tumors, some of the tumors weighing more than the test animal itself. Test animal after test animal developed cancer from the cancer tumor.

Rife then used the cancer MOR on the infected animals. Every one of the test animals recovered with no signs of cancer.

Next, Rife took the other half of the original tumor and subjected that tumor to the cancer MOR. He injected this into hundreds of rats. None of the rats developed any cancer. The cancer had been destroyed.

Dr. Walker at the University of California was so impressed with the results of Rife's tests on typhoid that he wrote..."if the ray should prove equally efficient in killing other pathogenic microorganisms, it would be the greatest discovery in the history of therapeutic medicine." [2]

A large experiment was conducted at the Scripts Ranch in La Jolla, California. The letters between committee members stated that 16 patients were treated for cancer and other major diseases. Within thirty days, 14 patients were diagnosed as clinically cured. The other two followed within 90 days.

In May of 1938, reporters from a local newspaper in San Diego came to Royal Rife and wanted to learn of the great discovery of a

---

[2] E. L. Walker letter written from University of California, to Dr. Milbank Johnson, October 18, 1935.

cure for cancer. Rife and his associates were very conservative in their response to the reporter.

"We do not wish at this time to claim that we have 'cured' cancer, or any other disease, for that matter. But we can say that these waves, or this 'ray', as the frequencies might be called, have been shown to possess the power of devitalizing disease organisms, of 'killing' them, when tuned to an exact, particular wave length, or frequency, for each different organism." [3]

Although there had been reports in many newspapers over the years regarding the incredible super Rife Microscope, this was the first time that a potential cancer cure was discussed.

Soon after this newspaper article, the members of the Special Reseach Committee began sharing that they were coming under pressure from several different directions. One medical society wanted to be a partner with Rife and sell the microscope. Rife did not want to do this. A perstigious family well-known in the United States was concerned about its pharmaceutical interests being affected and wanted information to be limited or stopped completely. A well-known group that was involved in cancer research told Rife and his group that they should never experiment with human cancers but stay exclusively to researching only cancers grown in animals.

The pressure began building to an extent that several doctors who had worked with Rife were threatened and many years later would still swear under oath that they never knew anything about any work with cancer, and they didn't know about any treatment device. This is very strange because I have many of their letters stating their exact relationship with Rife and his Rife Ray. I have several letters written in the 1930s in which a particular doctor wrote to Rife regarding his successes using the frequency treatment machine. I also have a letter written in the 1940s in which the same doctor wrote to the California Department of Public Health stating he never worked with Rife or knew anything about a microscope or treatment device.

In the early 1930s, there were quite a few newspaper articles regarding the microscope, but by 1940 almost no information was being reported. Rife found that many of his associates were creating distance between themselves and Rife's work. The Smithsonian article

---

[3] Newell Jones, *The Evening Tribune*, Copyright 1938.

appeared in 1944 and was written as if Rife had just developed the microscope at that time instead of when it was developed in 1931.

This lack of information continued for many years. As Christopher Bird stated in his article, *What has become of the Rife Microscope*, written in March 1976, "Calls to the U.S. Armed Forces Institute of Pathology Medical Museum, which has hundreds of different microscopes in its historical collection, to the National Library of Medicine's Historical Division, to the Smithsonian Institution and the Franklin Institute, both repositories for outstanding scientific inventions, and to a dozen establishments dealing daily in microscopy elicited from curators, medical pathologists, physicians and other scientific specialists only the complaint that none of them had ever heard of Royal Raymond Rife and his microscope." [4]

As most of Rife's supporters either abandoned him or died he became very depressed and eventually became an alcoholic. Years later while selling his special drafting set to get money to buy liquor, Rife would meet John Crane who sobered him up and heard the incredible story of his work back in the 1930s.

This is a sad story for many reasons when you think of how many people have died during the last 70 years from medical conditions that might have been addressed in a more effective manner. The war on cancer has been a colossal failure when you realize how much money has been spent and how few people really survive. When you read newspapers or watch television, you know that for all intents and purposes cancer is a death sentence. The medical doctors working with Rife wrote letters from their hearts and their experience sharing what they believed was a new hope for the world. These were the greatest researchers in the United States and belonged to the most prestigious institutions of medicine. They knew what they discovered, and they completely proved their discoveries. Out of all of the many discoveries I have made throughout the years of my journey, this story actually made me cry.

I now had in my possession this incredible Rife microscope. Occasionally we had it on exhibit at the World Research Foundation in Sherman Oaks, California. Because we had the microscope, it gave us a tremendous amount of credibility and many people became interested in our organization.

---

[4] Christopher Bird, What Has Become of the Rife Microscope, New Age Journal, March 1976.

In December of 1986, I was asked to do a one-hour radio program on KABC radio in Los Angeles, California. The *Open Mind* program was one of the top rated radio programs in the United States. This took me by surprise because I had hoped that when we would start discussing the microscope in public that my organization would have a strong and credible image. My organization was only two years old and not well-known.

There has always been some danger associated with the microscope over the years. There were threats made to Rife's group and I learned that several other people who had attempted to publicize the microscope had also had troubling experiences. You would not assume that people opposed to Rife in the 1930s and 1940s would still be around in 1986; however, people who were affiliated with the institutes where the Rife microscope had been used would have to explain how a 50,000-power microscope disappeared.

Some of the publicity included:

A full report of the Rife Microscope made in the Staff Meetings of the Mayo Clinic on July 13, 1932.

A dinner party honoring Rife and his research group was reported in the Los Angeles Times on November 21, 1931, and included a picture of the attendees who represented several major universities.

The New York Times reported Dr. Rife's work on November 22, 1931.

Science Newsletter made a report on December 12, 1931.

An article appeared in the Los Angeles Times Sunday Magazine on December 27, 1931. The subject of the article was the "Wonder-Work" of medicine in 1931.

The Christian Science Monitor reported on Rife's work on April 12, 1934.

The San Diego Sun reported the "World's Biggest Microscope" on April 16, 1934.

In these early days, stories and discoveries would be heard on one day and then might be forgotten or overlooked because nobody mentioned the subject again. We did not have any great archival system at that time. In the 1930s, we did not have news programs and talk shows similar to what we have today that can reach millions of people overnight. During Rife's era, the internet did not exist. Today over the internet, we can reach tens of millions of people in an instant.

I would have declined the invitation to be interviewed on the program; but Bill Jenkins, the host of the Open Mind show, told me that if I did not do the show at that time, he would interview John Crane. This was also a concern for me. Although John was a very nice man when I met him, he had a very unstable manner. If you knew his story, it might not be so surprising that he acted the way he did. Because of what Rife shared with Crane regarding his therapy device, Crane felt it important to make the information public. Crane attempted to construct and sell therapeutic devices based on Rife's work, and the result was that authorities in California arrested him. Crane's trial was held in California, and although Rife was still alive, he was not a part of the proceedings. My explanation as to why Rife was not included is that the authorities felt that Crane was an engineer and was an easy target where Rife would have opened up a more scientific aspect to the trial.

When John attempted to contact a few of the doctors who worked with Rife some 25 years earlier, he was not able to contact them. It appears that the prosecution in John's trial produced interviews on the Department of Health stationery from two prestigious doctors from Rife's group, stating that they never knew anything about work with cancer and that no one knew how to understand what was seen through the microscope. This is disgusting considering I have copies of letters, from the 1930s, in which the doctors remarked they "were dreaming" about the new discoveries that were being accomplished.

After being released from jail, John's bitterness left him very unstable, and he had lawsuits against every member of Congress, the President of the United States and several other prominent people in society who he believed had crushed this research work and placed him in jail.

I made the decision to do the program and make the presentation of Rife and his work. LaVerne and I did the radio program on a Saturday evening. A one-hour program became a three-hour program due to the overwhelming response of the callers to the show. After the radio program ended, we continued taking calls for another two hours.

This sudden exposure and increased interest in the microscope was to have an unfortunate result for me. Two weeks after the program, I was contacted by three lawyers stating they were representing John Crane and other "interests". They told me that John wanted his microscope returned. I responded as they requested. Later

I learned that a fight ensued among these people regarding who would possess this device. In 1987, these people each took different parts of the microscope. I have since learned that the microscope now safely exists in one place and with all its parts.

Two months after returning the microscope, I was contacted by Robert Maver, who was the head actuary for Mutual Benefit Life. Robert had heard the KABC interview and wanted to know how he could help. I had to explain to him that I no longer had the Rife Microscope. Maver asked if we could still meet so he could learn of any other interesting items our foundation might be aware of in the health field. Maver and I met for four days, and he used ten audio tapes for our meeting.

BACILLI REVEALED BY NEW MICROSCOPE

Dr. Rife's Apparatus, Magnifying 17,000 Times, Shows Germs Never Before Seen.

Special to The New York Times.

LOS ANGELES. Nov. 21.—A description of the world's most powerful microscope, recently perfected after fourteen years' effort by Dr. Royal Raymond Rife of San Diego, was one of the features of a dinner given last night to members of the medical profession by Dr. Milbank Johnson in honor of Dr. Rife and Dr. Arthur I. Kendall, head of the department of research bacteriology of the Medical School of Northwestern University.

The strongest microscope in use magnify 2,000 to 2,500 times. Dr. Rife, by a rearrangement of lenses and by introducing double quartz prisms and illuminating lights, has devised apparatus with a maximum magnification of 17,000 diameters.

Dr. Kendall told of cultivating the typhoid bacillus on his new 'medium K.' This bacillus is ordinarily nonfilterable. By the use of Dr. Rife's microscope, Dr. Kendall said, the typhoid bacilli can be seen in the filterable or formerly invisible stage.

New York Times, Sunday, Nov. 22, 1931

Robert Maver was so excited about the other items we had within our organization that he left from our meeting, flew to New Jersey, the main headquarters of Mutual Benefit Life. He met with the president of his company, and they created a special research division to investigate all of the information that he received from our meeting. World Research Foundation became the consultant for this special research division of Mutual Benefit Life. Robert Maver would later publish information we supplied to him to other members of the insurance industry.

It was because of the Rife Microscope that Mutual Benefit Life was attracted to our organization and their support allowed us to expand our operations and assistance to the public. Therefore, I can not say that for me "nothing happened."

The story of Royal R. Rife and the development of his universal microscope is a story demonstrating how creative genius should not have to accept preconceived limitations. In the case of Rife's microscope, the limitation is one that scientists formulated as a theoretical law of light and magnification. Rife did not accept this as fact and far surpassed the law that still hampers our scientists today! Using my earlier analogy of investigating a new continent and only be allowed to travel north, Rife traveled in the direction his inner compass pointed.

The Rife story is also surrounded by greed and suppression and demonstrates some of the lowest aspects of humankind. In the 1930s, high-ranking officials within the medical establishment tried to pressure Rife to assign the rights to his discoveries to them. These were people who were never involved with his research and never provided any research funding. When Rife refused to go along with the wishes of some of our "bastions of medicine," they made a point not to inform other doctors and researchers about his incredible device, Thereby cheating researchers of his development of a super microscope that opened up new worlds never witnessed before. The Rife Microscope would have provided researchers with a deeper view within many different medical and scientific fields.

Furthermore, when Rife and his associates were able to demonstrate a new therapeutic approach that was based on solid aspects of physics, they upset another group of people who were working hard to convince the world that the only approach to medicine was through chemicals. The pharmaceutical industry would not accept anything but its chemical approach as the medical therapy of choice. The pharmaceutical companies would later influence legislators to pass laws restricting the use of many of the other existing medical approaches.

Rife and his associates were told that they should only experiment with cancer induced in animals and not work with human cancer that was implanted into animals. I want this point to become very clear for you. In his early experiments, Rife was not treating cancer in people but taking their cancer tumors, placing the tumors inside the animals, and then curing the animals of the human cancer. Rife was told to stop researching in this manner and only work on non-human cancer grown in animals. Rife and his group could not believe this and neither can I. Isn't the whole goal of cancer research the curing of human cancer?

Through use of the Rife Microscope, Dr. Kendall of Northwestern University was able to prove that the theory of pleomorphism was a reality. Pleomorphism is a theory that says that microorganisms in the body can mutate from one state to another and could, therefore, be the cause of many of our health illnesses and diseases. This means that when the environment within our bodies changes, a non-threatening microbe or germ can mutate into a deadly pathogen and become the cause of our health problems. This also means that not all of our health problems come from someone we come in contact with but because of what is going on within our own bodies. Dr. Kendall and Rife were able to prove this theory because they could watch it taking place **live** under Rife's super-microscope. Rife's group discovered ten basic groups of germs that could mutate from one form to another within their basic group. Other medical researchers did not have the microscope, and, therefore, did not believe it. This concept is not acceptable to the pharmaceutical industry, which was pushing inoculation therapy with great gusto. Vaccination may have a place for some health problems, but is not the solution for all health problems. This would explain why diet and nutrition are so important to keep a healthy environment within your body. Many years ago, the government and non-profit organizations for cancer research laughed at the concept that diet, nutrition, minerals and vitamins had any affect upon cancer. These groups also scoffed at the idea that emotions could have an effect on cancer.

The greatest use of the powerful Rife Microscope would be to view drug resistant bacteria, germs or other microorganisms and find

UNIVERSITY OF CALIFORNIA

THE GEORGE WILLIAMS HOOPER FOUNDATION

SECOND AND PARNASSUS AVENUES

SAN FRANCISCO

March 6, 1934.

Dr. R. R. Ryfe,
708 Electric Building,
San Diego, Calif.

Dear Dr. Ryfe:

I am still "dreaming" about the many things
you were kind enough to show me last Saturday. As soon as I
can tear myself loose I will accept the privilege of coming back
and bringing with me some of the agents which produce disease.
The tumor which I brought with me in the two rats is Hyde 256
carcinoma. I hope it will be of some use to you.

With kindest regards and best wishes, I am,

Sincerely yours,

K. F. Meyer

53

2. Keith, H. M.: The effect of various factors on experimentally produced convulsions. Am. Jour. Dis. Child. *41:* 532-543 (March) 1931.
3. Keith, H. M.: Further studies of the control of experimentally produced convulsions. Jour. Pharmacol. and Exper. Therap. *44:* 449-455 (April) 1932.
4. Keith, H. M.: Factors influencing experimentally produced convulsions. (In press)

OBSERVATIONS ON FILTER-PASSING FORMS OF EBERTHELLA TYPHI (BACILLUS TYPHOSUS) AND OF THE STREPTOCOCCUS FROM POLIOMYELITIS

E. C. Rosenow, M. D., Division of Experimental Bacteriology: Interest in the series of reports by Dr. Kendall[3,4] on the filtrability of bacteria by the use of his protein-rich, peptone-poor, K medium reached its climax, it might be said, following publication of the report on the successful visualization, with the high-power Rife microscope, of the filter-passing form of Eberthella typhi in cultures of K medium, and in the corresponding filtrates.[5] Discussion over these important, and in some respects revolutionary, findings has become widespread. Since I, as well as others[5,6,7,8] have been attempting, and with some success, to make the streptococcus, especially that from poliomyelitis, filtrable and to develop the virus of poliomyelitis from the streptococcus so consistently isolated in this disease, I accepted a recent kind invitation of Drs. Kendall and Rife to share with them their observations in a restudy of the filter-passing forms of Eberthella typhi as seen with an improved model of the Rife microscope. They asked me also to bring with me my cultures of the streptococcus from poliomyelitis.

It is the purpose of the report to record the more important observations made during three days, July 5, 6, and 7, 1932, spent with them in Dr. Kendall's laboratory at Northwestern University Medical School, Chicago. Owing to the novel and important character of the work, each of us verified at every step the results obtained. Microscopic examination of suitable specimens was made as a routine by Dr. Rife with his high-power microscope, by Dr. Kendall with the oil immersion dark field, and by myself with the ordinary Zeiss microscope equipped with a 2 mm. apochromatic oil immersion lens and x10 ocular giving a magnification of about 900 diameters. Most observations with the Rife microscope were made at 8,000 diameters. In order to check the magnification, gram and safranin stained films of cultures of Eberthella typhi, of the streptococcus from poliomyelitis, and stained films of blood, and of the sediment of the spinal fluid from a case of acute poliomyelitis, were examined. Bacilli, streptococci, erythrocytes, polymorphonuclear leukocytes, and lymphocytes were clearly seen, and in each instance were, as near as could be estimated, about nine times the diameter as when examined with the 2 mm. oil immersion at about 900 diameters.

The following principles and methods were stated by Dr. Rife as being essential in order to visualize clearly the objects at this and higher magnification by direct observation. Spherical aberration is reduced to the minimum and magnification greatly increased by using objectives in place of oculars. Proper visualization, especially of unstained objects, is obtained by the use of an intense beam of monochromatic polarized light out the life frequency of those organisms. Each time a drug resistant bacteria mutates it has a new slightly altered life frequency or MOR that we would use to destroy it. Imagine the implications!

Research organizations keep telling us that the answer is right around the corner if we just keep giving them more money. The truth is that the answer is right in front of us. Rife said in his last words before he died, "…the most important thing I ever did was build a microscope."

Rife did his part. He built a super-microscope for all of us - and nothing happened!

Some Benefits of using the Rife Microscope:
- Determine the frequency to destroy any microorganism that is the cause of a current, new or unknown medical condition.
- Determine the new frequency of drug resistant bacteria that mutates. As it mutates there would be a new frequency that would destroy it.
- Examine the cells of a patient being treated. The true effectiveness of any therapy used on a specific patient can be determined.

Some Benefits of using Frequency Rays:
- Destroy microorganisms that are the cause of 50 illnesses and diseases.
- No chemical adverse effects that might lead to other health problems.
- Application of frequencies can change harmful microorganisms into less harmful activity.

ONLY YOU CAN MAKE IT HAPPEN!

"Narrow mindedness cannot follow upon broad viewpoints."

Chapter 3
The Nobel Chairman Is Ignored

Even though as I've confessed I don't possess any mechanical ability that does not prevent me from having a great interest in people who have that talent. Being mechanically challenged, I have always been cautious of going near something that was connected to a power outlet or fuse box. I make sure to check with some expert, or at least someone wiser than I, before venturing onward in this area. However, I have always had a great fascination with electricity and magnetism.

I assume most people are fascinated by these elements. With magnetism, there is something interesting about particles jumping through the air to stick tightly on a magnet. When you attempt to push two magnets together of the same polarity, you can feel the force even though you don't see anything in the space between them. When you are playing with two magnets with very strong strength, you can feel that incredible, increased unseen force.

Who is not awed when watching electrical storms and bolts of lightning shooting through the sky and illuminating the darkness at night? Growing up in the Los Angeles area, we did not have many electrical storms, but they were very impressive when they did happen. Living in Sedona I am treated to a greater number of storms, and the lightning bolts seem to shoot in many directions as they explode against the sky. Many times, I've been watching a storm, and I have seen the lightning travel sideways rather than touching the ground. Often we have dry lightning storms that suddenly appear. While these storms are beautiful to watch when you are at a safe distance, they do cause many fires in our area. Another problem we have is more than our share of power outages when the strikes hit too close to transformers.

For most of us growing up in the United States, we are familiar with the numerous movies and television shows about Frankenstein and his monster. Great bolts of electricity, screaming and blasting at our senses, are pumped into Frankenstein's creation bringing it to life. Although these are fictional stories, the truth of the matter is that without electricity, we would not have life. Every one has seen enough television programs and movies to know that when the heart stops, the paddles of a difibulator are used to restart the heart. We are, after all, an electrical device. What determines life? In medicine, the determination of life is a heartbeat or electrical brain waves. A person is considered dead if there is no perceptible electrical current in either the heart or the brain.

Electricity is an amazing force. When a person is electrocuted he or she is touching the same force that courses within the body, but that force short-circuits the system and the person dies. Overall, I'm sure that if you ask most people about electricity they consider it a force around which they should be most cautious.

In my first sixteen years of research in the health field, I accumulated a good deal of information pertaining to the use of magnetism and electricity in medicine. The majority of the information was from researchers in the 1930s through 1950s.

However, in 1986 I had the opportunity to learn about some exciting research involving the use of electricity to fight cancer. During that year, my associate LaVerne and I thought that it would be beneficial if we would hold a congress where we would bring some of the greatest medical doctors and scientists in from around the world.

Our World Research Foundation was only two years old, and we thought it would be of great interest to bring together various people who were doing work in the electromagnetic spectrum. The electromagnetic spectrum includes all frequencies from zero to infinity. The only way any of us learn is through investigation, research and exposure to new ideas and thoughts. We were not promoting any particular therapy or technique but bringing people together for discussions of what was known at that time. Our discussions were to include the areas of magnetism, electricity, sound, color, music & light.

One of the medical doctors who had already agreed to speak at our 1986 congress suggested that I contact Dr. Bjorn Nordenstrom from Sweden. Dr. Nordenstrom was creating quite a stir in scientific circles because of his work in shrinking lung and breast cancer tumors using electricity.

The use of the words cancer and electricity immediately tweaked my interest and curiosity.

Dr. Nordenstrom had proved that there is electrical activity taking place within the body, and it is the very foundation of the healing process. Nordenstrom had been making the point that electricity in the body is as critical to well-being as the flow of blood.

Doing a little bit of name-dropping allowed me to speak directly with Dr. Nordenstrom. I reached him at the university at Karolinska Institute and invited him to be a speaker at our world congress. He accepted the invitation and promised to send information regarding his background and accomplishments.

What immediately caught my attention is that Nordenstrom was a member of the Nobel Assembly since 1967, President-Elect of the Nobel Assembly in 1984 and President of the Nobel Assembly in 1985. His list of academic appointments, hospital appointments, education and list of more than 122 published papers was also quite impressive. One would certainly assume Nordenstrom would have great credentials considering he was the Chairman of the Nobel Assembly of Medicine!

He had also written a book titled, *Biologically Closed Electric Circuits*, which provided information pertaining to his theory of the bio-electromagnetic energy of the body.

I was curious how someone might attain the lofty position that Dr. Nordenstrom had attained. How does one become the head of the Nobel Assembly, which selects the Nobel Laureates in medicine? I discovered that Nordenstrom had pioneered what is now known as percutaneous needle biopsy, a diagnostic technique used in every major hospital in the world. Who has not heard of needle biopsy?

Apparently, when Nordenstrom first proposed needle biopsy it met with some heavy resistance. Many medical specialists believed that this technique would not be possible, and that it would be dangerous. Nordenstrom proved that it was possible, not dangerous, and his persistence led to it becoming a standard diagnostic tool utilized around the world.

Nordenstrom's background and credentials showed that he was definitely an innovator. Based on his previous record of accomplishments, one would assume that when he placed forth a new theory that it might have easier acceptance in the medical community.

His current theory regarding the electrical influences upon the human body was meeting with some resistance. Nordenstrom's discovery would cause a completely new thought about how the body operates.

Nordenstrom's theory is that an injury or tumor alters the normal flow of the bio-electromagnetic energy that is contained within the closed circuit network of tissue, arteries and finally the veins. Repairing the closed circuit that has been altered would lead to helping the body recharge its healing power.

In some of the following chapters of this book, you will learn of other prominent medical and scientific experts in the 1930s and 1940s who proved the influence of electricity on life. To state it more exactly, you do not have biological life without electricity. When any biological activity or movement takes place within the body, a measurable electrical action takes place. In later chapters, I will introduce researchers who identified and measured these activities.

Dr. Bjorn Nordenstrom

Research demonstrates that there is not a small amount of electrical action, but a universe of electrical activity that takes place in the human body. You are aware that we have veins and arteries that carry the blood throughout our system; imagine that there are similar electric circuits within the body. These circuits resemble the electrical lines that you see around your home or office. Although there is a constant flow of electricity within our bodies due to the normal activity of internal organs and functions, the electric circuits are more active when we sustain an injury, incur an infection or when a tumor is present.

Voltages build and fluctuate; electric currents course through arteries and veins and across capillary walls, drawing white blood cells and metabolic compounds into and out of surrounding tissues. This electrical system works to balance the activity of internal organs and in the case of injuries represents the very foundation of the healing process.

The most profound aspect of Nordenstrom's theory is that disturbances in our electric network may be involved in the development of cancer and other diseases. Nordenstrom believed this truth because he was able to measure changes in voltage at the site of a tumor as it is increasing or decreasing.

Dr. Nordenstrom's approach to cancer seems logical when you consider his theory. With cancer patients, he would insert small electrodes into the cancer tumors as well as in the nearby healthy tissue, thus redirecting the body's bioelectric energy. He would place the positive electrode in the tumor while placing the negative electrode into the normal surrounding tissue. Using the concept of opposite charges attract, Nordenstrom induced the tumor-fighting white cells to the tumor site.

As a tumor grows, the inner cells of the tumor are cut off from the normal circulatory system and slowly die. Because of the cell death there are chemical changes in the tumor; what occurs is the building up of a larger positive potential in the tumor.

By running a current into the tumor, Nordenstrom prolonged the electro-positive phase in the circuit and this triggers a variety of tumor-fighting effects in the body. The white cells are attracted and begin working on the tumor. The best example I can think is for you to visualize the Pac-Man game. The white cells begin chewing up the tumor to get to the probe. Opposite polarities attract and those cells attempt to reach the probe.

This is a cursory explanation of a technique that Nordenstrom spent more than forty years perfecting.

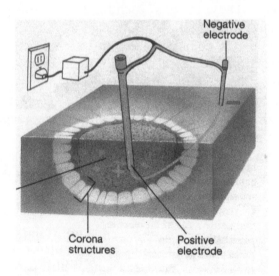

Probe placed in tumor, Discover Magazine, April 1986

I had the opportunity to spend a week with Dr. Nordenstrom while he was at our World Research Foundation congress. Nordenstrom made two presentations, a one-hour lecture and a two-hour seminar. Videos taken at the congress are rare records of Nordenstrom presenting his work.

While Nordenstrom was at our congress in California, the news program 20/20 called to schedule an interview with him. When I told Dr. Nordenstrom about the program's request, he was going to decline the interview. Dr. Nordenstrom had little patience with the media. He shared that virtually every interview that was conducted by an American news service had been highly critical, skeptical and at times demeaning. The media were saying that Nordenstrom was dealing in *alternative* medicine. The only publication that presented a factual and fair account was an article that appeared in the April, 1986, issue of <u>Discover</u> <u>Magazine</u>. He considered it a fair presentation of his work.

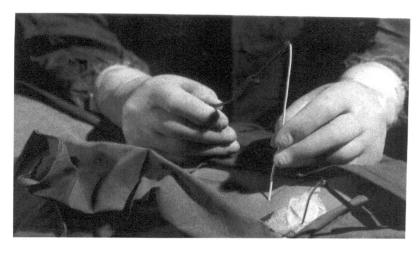

Dr. Nordenstrom inserts probe into cancer tumor

Dr. Nordenstrom related his tremendous displeasure at an article that appeared in the Los Angeles Times that made his work seem very trite and "way out"! Nordenstrom preferred gatherings that were more scientific where he could present his research to colleagues who would take the time to listen to his data. As an example, he enjoyed his time speaking to medical doctors at the City of Hope Hospital in Los Angeles. The administrators of that hospital contacted me before

61

Nordenstrom arrived and asked for permission to invite him to make presentations.

I convinced Dr. Nordenstrom that the 20/20 program would be to his benefit and he finally agreed to do the interview. The program aired nationally exactly two years later in 1988. [1] The responses of the medical doctors interviewed on the 20/20 program were much the same as responses coming from other medical specialists. Briefly, they thought Nordenstrom's work was very interesting, but they did not want to take the time to look at something that was different from the training they had received. Nordenstrom's material brought them into the area of physics, and they were completely lost in the ramifications.

The resistance was the same as when he proposed the needle biopsy. Nordenstrom told me that when he brought his book to the medical book publishers they had turned it down because they did not believe enough doctors would be interested in purchasing it. If they weren't going to have sales they wouldn't bother publishing it. Therefore, Nordenstrom published the book himself. However, because the book was self-published, some medical authorities stated that it did not have the same credibility as books printed and distributed by recognized medical publishers. It was the old catch-22.

In addition to the media, some people in the medical community were labeling Dr. Nordenstrom as an alternative health practitioner. During a question and answer session after his presentation at our congress, someone asked him if he considered himself an alternative medical practitioner. He answered with complete disgust at the question. The individual who posed the question remarked that under current laws and regulations in many states, a therapy used for cancer that is not chemotherapy, radiation or surgery is alternative and therefore might be illegal.

In the few articles that objectively reviewed Dr. Nordenstrom's work, some of the reporters mentioned that his work was perhaps the most profound biomedical discovery of the century. I know that Nordenstrom basked in that appellation.

Regarding the idea of it being the discovery of the century, I must share the following event that took place. Before Dr. Nordenstrom left to return to Sweden, he came to the World Research Foundation

---

[1] 20/20 program aired on weekend of October 21-23, 1988

offices to present an autographed copy of his book, *Biologically Closed Electric Circuits,* for our library.

In a ceremony befitting the stature of Dr. Nordenstrom, we placed his autographed book in the "electricity" section of our library. As Nordenstrom placed his book alphabetically on the shelf, he noticed a little black book next to his.

This book titled *The Application of Electricity as a Therapeutic Agent* was written in 1877. [2]

Glancing at the book, Dr. Nordenstrom discovered that a medical doctor had utilized electricity to shrink a patient's breast and lung cancer tumors. Of course, Nordenstrom noted that this book was more than one hundred years old.

I do not want to take anything away from the brilliance and discoveries of Dr. Nordenstrom. Obviously, he was not the first discoverer of this therapeutic possibility. The doctor who wrote about it in 1877 did not understand why the therapy worked. Dr. Nordenstrom had a greater understanding of what was taking place in the body, and he should be acknowledged for what he accomplished.

Through the 1980s and 1990s, Nordenstrom's work would find little support or acceptance in the United States. Only medical doctors in the People's Republic of China showed real interest in his medical approach to cancer. Nordenstrom traveled to China where he established medical centers to teach his technique. Nordenstrom placed his patents, free of charge, in public domain in China to allow others to continue his work. Nordenstrom passed away without seeing his theories accepted as a part of the mainstream medicine in the United States.

While the medical community in the United States was showing no interest, it is obvious that the medical doctors in China believed there was something of value in his work. Since his first visit to China, more than 2,300 Chinese doctors treated more than 13,000 tumor patients with Nordenstrom's technique with an effective rate of 75%.

Based on the incredible success of his therapy, in 2002 the government of the People's Republic of China awarded Dr. Bjorn Nordenstrom the highest scientific award that a foreigner can receive -

---

[2] J. H. Rae, *The Application of Electricity as a Therapeutic Agent*, (Philadelphia: Boericke & Tafel, 1877).

The International Scientific and Technological Cooperation Award of China. [3]

Nordenstrom tried to share his great discovery with the U.S. medical community - and nothing happened.

Some Benefits of Nordenstrom's Therapy:

- No adverse effects from treatment such as loss of hair or stomach sickness.
- Patients keeps their lungs or breasts.
- No secondary tumors caused through radiation and chemotherapy.
- Treatment costs are significantly lower than standard drugs.
- Patients do not pass harmful chemicals through their urine into water supplies.

ONLY YOU CAN MAKE IT HAPPEN!

---

[3] Embassy Of People's Republic of China in Sweden, "Swedish Expert Awarded the International Scientific and Technological Cooperation Award of China", March 11, 2002

"There is not one blade of grass; there is no color in this world that is not intended to make us rejoice." John Calvin

Chapter 4
Who Is Colorblind?

One morning during April of 1977, I awoke remembering a profound dream that I had during the night. In the dream, thousands of sick people were either sitting or lying in beds in a large multiple storied building. I heard beautiful music playing and all of the people began walking out of the building. As the people went outside, they noticed a giant rainbow in the sky even though there were no clouds. The rainbow colors began bathing all of the people, and they immediately were in perfect health.

I awoke having a belief in a new manner of healing. Sharing this concept with my father elicited the response, "Even if it is true, how will you ever figure out which color would be helpful for different health problems?"

Although my father's point was certainly logical, I did not have any doubt that somewhere I would find an answer. Actually, my quest was far easier than I would have ever imagined. When I arrived at my favorite bookstore and asked if there were any books written about the use of color and healing, I discovered there was an entire section of books dealing with color therapy.

The first book that caught my attention was Linda Clark's book, *The Ancient Art of Color Therapy*. This book proved to be a wealth of information, especially reading that color therapy was an ancient healing practice.

Linda Clark's book provided information on several medical doctors and health professionals who had been using color therapy with their patients. The manner in which the color is used seemed both simplistic and amazingly simple. Starting with Clark's book and trusting that my dream was inspiring me to look further, I began researching the possible use of color in the healing process.

Whenever I begin researching a new potential health therapy, my first concern is whether the therapy works and if it is safe. My initial research is not based on some expert's opinion regarding the therapy but on someone who has been utilizing the therapy or technique. There is an important distinction here. Often I find, similar to my experience in college with the sports physician whom I consulted for my injury, people will make comments about something that they themselves have never experienced.

My good friend Professor Dr. Karl Walter once told me that he had met many a fellow university professor in Germany who made the statement, "must not work because it can not work." His point was that even when confronted with some fact, many people disregard their own senses because their minds think something is impossible. Karl met scientists and physicists who would see something taking place right before their eyes, but they would deny that it was happening because it contradicted their own beliefs. Perhaps this seems a confusion of words, but it is very true for some people in academia. This is the dilemma of the sacred cow. If experts accept a new fact that contradicts a point of their very own theory, then their credibility comes into question. Therefore, they make comments about something they might never have examined on their own.

When I am giving lectures and I refer to color, often a medical doctor in the audience will ask if the use of color is real and could it possibly be beneficial. Often the question is made in a very sarcastic manner. I will pose the following question to the medical doctor: "What is the therapy for babies born prematurely with infant liver syndrome? I might add, doctor, what would you use for the condition called Crigler Najjar Syndrome (CNS) (yellow jaundice)?" The doctor will respond: We use a blue." Then his voice tails off. I then say: "That is right, you use a blue light or the baby with this condition will die! Doctor, I do not know what you call that, but I would call it color therapy."

In the late 1950s, a nurse in Great Britain first discovered the value of sunlight in alleviating jaundice in newborn babies. The story has circulated that the babies near the windows lived while the babies inside the rooms would die. Subsequent study showed that it was not the total light but the blue component in the sunlight. Blue light phototherapy (another word for color therapy) is the therapy of choice to keep babies alive in their beginning stages of life. In the early 1970s when CNS was first reported, most of the babies died in infancy.

The blue light contacts the blood vessels and tissue close to the skin, and the bilirubin undergoes a number of changes. The sequence that takes place under the blue light is slightly different than the normal way that the body functions when bilirubin is present. This

66

color therapy is the only thing that helps prevent the CNS patient from sustaining brain damage.

Now is a good time to explain color therapy. Color therapy is the use of the colors of the visible spectrum. Think of the various colors that you see in a rainbow or through a prism. Each color is determined by a rate of vibration. As vibrations change the color you see also changes. In color therapy, a color is projected over a distance onto bare skin at particular parts of the body. The location on the body is determined by what the particular health difficulty might entail. There are specific colors or combinations of hues used for specific ailments or health problems. The color affects the body in a manner that allows the body's own healing abilities to become more effective. It is not that a specific color is curing a specific health problem; the color is causing physiological change within the body that allows the body to be operating in a more efficient manner.

During the course of the last fifteen years, several popular magazines and even a few television programs have noted the effect of various colors on our moods. An article actually appeared in Reader's Digest many years ago relating how the color of the walls where children studied could effect their IQ. Reports have shown that when a violent prisoner is placed in a jail cell with pink walls in a matter of minutes the prisoner becomes very docile. In fact, when a poster board painted pink is placed at eye level in front of a weight lifter, the weight lifter's strength becomes greatly diminished.

I had discovered that the color blue was saving the life of sick babies, that the color pink could rob a weight lifter of his strength. Linda Clark's book had mentioned that doctors could cure cataract with color, so I wondered what other medical conditions could be helped. I found the answer to my question when I learned of the work of Dinshah Ghadiali and his color therapy system called SpectroChrome.

My first contact with Ghadiali came when I found a sequence of photographs taken in 1926 of a little girl with 3rd degree burns over 2/3rds of her body.

As you will see, the following pictures are so dramatic, and the case was so compelling, that it is impossible to ignore. The pictures were taken out of the book, *Let There Be Light*, by Darius Dinshah.

Dinshah

67

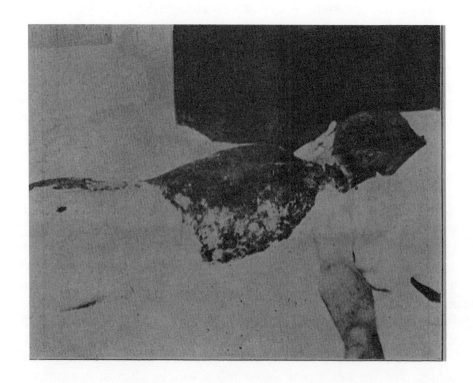

Grace Shirlow 2 weeks after admission

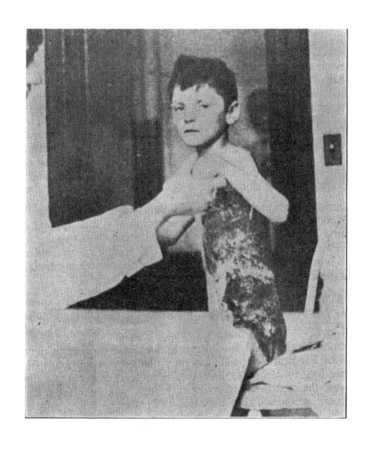

Grace Shirlow 2 weeks after admission

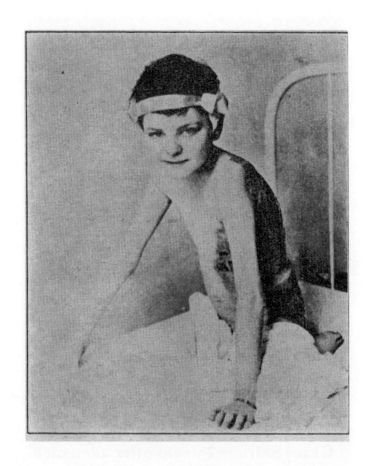

Grace Shirlow 3 months after admission

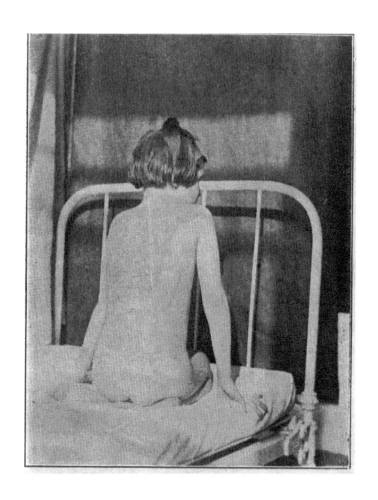

Grace Shirlow 18 months after fire

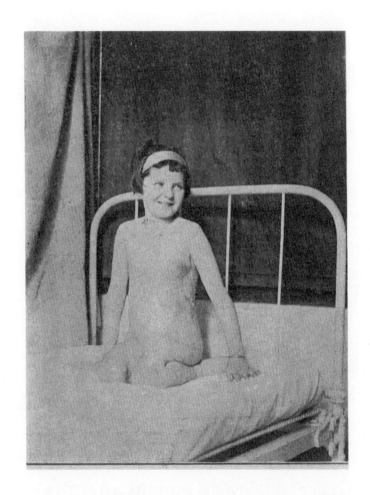

Grace Shirlow 18 months after fire

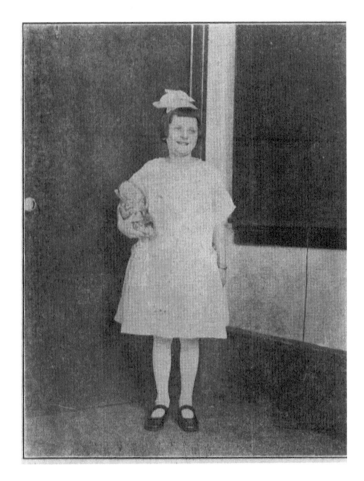

Grace Shirlow leaves hospital

The little girl in the pictures is Grace Shirlow. In 1926, she arrived at Philadelphia Women's Hospital where the majority of the staff of the hospital considered her a hopeless case. Eight-year-old Grace had been burned badly on much of her body; she had a body temperature of between 105 to 106 degrees and had almost complete suppression of urine for more than 48 hours.

Fluids were forced into her, but with no effect. Only one medical doctor on the staff had the belief that she could help this poor child. Dr. Kate Baldwin, who was the Senior Surgeon at the hospital, was not on call when the girl arrived but reported to the hospital several hours later. When she examined the girl and everyone present had given up hope, Dr. Baldwin proposed the use of color therapy. For several years before this case, Dr. Baldwin had been using a color therapy system called Spectro-Chrome, a method of color therapy using 12 specific colors.

Dr. Baldwin used the color scarlet over the girl's kidney, and within minutes, the little girl voided more than 6 ounces of fluids. Dr. Baldwin then projected the color blue over Grace's body and within one hour her incredible pain had subsided. In fact for the duration of Grace's stay in the hospital she did not experience any more pain from her burns. For me this is unbelievable! Think about the pain and suffering that all burn victims experience. How can anybody justify not looking into this approach?

The medical report of this case stated that no topical application was made, and no skin grafts were performed. Grace Shirlow grew new skin during the course of her Spectro-Chrome color therapy. Although this is a very dramatic burn case, it is not unique. A full report of the Grace Shirlow case was delivered at a medical meeting[1] and written up in a medical journal. [2]

The inventor of the Spectro-Chrome system is Dinshah Ghadiali. Dinshah was born in 1873, and he entered high school in 1884. This is not a typo; he entered high school at the age of eleven. Dinshah took his entry examinations for Bombay University at the age of thirteen. Dinshah made his first visit to the United States in 1896 and had the opportunity to meet Thomas Edison and Nikola Tesla, two of the most famous inventors and scientists in the history of our planet. Dinshah

---

[1] Dr. Kate Baldwin, Section on Eye, Ear, Nose, Throat Diseases of the Medical Society of the State of Pennsylvania, Oct. 12, 1926.
[2] Dr. Kate Baldwin, *Atlantic Medical Journal*, April 1927.

gave lectures on x-rays and radioactivity and the New York Times termed him the "Parsee Edison".

Dinshah and his wife and family immigrated to the United States in 1911, and Dinshah became a naturalized citizen in 1917. In 1919, Dinshah was appointed Governor of the New York City Police Aviation School and later was commissioned Colonel and Commander of the New York Police Reserve Air Service. Two aircraft obtained from the U.S. Government were used to patrol the harbor in the city of New York. Dinshah flew the first police airmail service from New York to Philadelphia. For his meritorious service to the city, New York Mayor John Hyland awarded Dinshah the Liberty Medal.

Dinshah delivered his first lecture on his Spectro-Chrome system in April of 1920. Dinshah would eventually present more than one hundred lectures throughout the United States.

At one time nearly 500 United States medical doctors were using his Spectro-Chrome system. Although the medical doctors, like Dr. Kate Baldwin, were achieving results that were remarkable, Dinshah Ghadiali was brought to trial in Camden, New Jersey, in 1931, for a charge of grand larceny. The complaint from the medical authorities was that Dinshah was committing fraud because Spectro-Chrome could not have any effect on diseases.

During his trial in 1931, Dinshah called upon three witnesses. The first witness was Dr. Kate Baldwin. Dr. Baldwin testified under oath that she had been a medical physician for 40 years and senior surgeon of Philadelphia Women's Hospital for 23 years. Under cross examination by prosecutors Dr. Baldwin stated that she had successfully used the Spectro-Chrome system for; cataracts, glaucoma, hemorrhage in the retina, ordinary inflammatory conditions of the eye, sclera, infection of the sinuses, bronchitis, pneumonia, pleurisy, tuberculosis, ulcers of the stomach, asthma, jaundice, kidney conditions, appendicitis, gonorrhea, syphilis, breast tumors and severe third-degree burns. Dr. Baldwin said that she was using up to ten Spectro-Chrome light machines in her medical practice.

Dr. Martha Peebles appeared as a second witness for Dinshah. Dr. Peebles was a doctor of medicine for more than twenty-four years and had worked for the Department of Health in New York City for over twenty years. Dr. Peebles was a physician to the American Expeditionary Forces during World War I. During the war, she would perform up to 61 different surgeries a day.

75

Dr. Peebles was forced to retire from active practice due to ill health, but when another doctor had used the Spectro-Chrome system on her, it restored her to health. At the time of the trial, Dr. Peebles was using 17 Spectro-Chrome machines in her practice. Dr. Peebles swore that she had treated cancer, arthritis, poliomyelitis, mastoiditis and many other medical conditions.

Dr. Welcome Manor, M.D., was the third doctor who appeared as a witness for Dinshah. Dr. Manor had practiced medicine for more than thirty years and testified that he had successfully treated cancer, diabetes, gonorrhea, syphilis, ulcers, hemorrhage, neuritis, spinal meningitis, heart disorders and numerous other maladies.

The jury in this 1931 case was out only 90 minutes and returned a verdict of not guilty for Dinshah.

However, Dinshah's legal woes were not over. He was brought to trial for one charge after another and served a combined 18 months in jail.

In 1947, the FDA instituted a trial against Dinshah for practicing medicine without a license. Dinshah lost the case, and this effectively forced him to surrender all of his research equipment and books. According to the judge's decree, all of Dinshah's books were destroyed and Dinshah was placed under probation for five years and not allowed to discuss the Spectro-Chrome form of therapy. I've seen photographs of agents burning his books. It is heartbreaking.

In 1958, the FDA obtained a permanent injunction against Dinshah preventing any machines or books from crossing state lines. Dinshah passed away in 1966.

In 1924, in the section titled "Minutes of the Journal of the American Medical Association," an article appeared that completely discredited Dinshah as being silly and suggested that if his therapy really worked why was he wearing glasses. The article pointed out that any doctors following this system would be scrutinized for following such a stupid therapeutic approach to medicine.

Dr. Kate Baldwin was confronted by other personnel at her hospital and the Board of Directors gave her this following choice: "Either give up your use of the Spectro-Chrome machines or leave the hospital and your prestigious position." Dr. Baldwin stated that she would give up the practice of medicine rather than stop using a therapy system that was better than any other system available to a medical doctor. She resigned her position at the oldest surgical hospital in the United States.

During the years, there have been several hundred studies and reports regarding the use of color in the healing arts. In 1981 at the School of Nursing, College of Human Services, San Diego State, California, Sharon McDonald, Ph.D., conducted an interesting study of the effects of visible light waves on arthritis pain. In her controlled study she worked with 60 female volunteers between the ages of 40 and 60 years old who had confirmed medical diagnoses of rheumatoid arthritis. Dr. McDonald constructed a box that would allow her volunteers to place their elbows inside, and then she would project either a red light, blue light or no light upon their elbows. The volunteers did not know which if, any color, was used. Dr. McDonald discovered that the longer the blue light was used the greater reduction in pain. When no color was used, there was no influence in reduction or increase in the pain felt by the volunteers.

I have delivered several lectures as well as articles dealing with the use of color therapy. In a medical magazine dealing with cancer research for clinicians, an article appeared listing my organization as pushing the ridiculous concept that color could work for health problems. The article, which appeared without an author, stated that no reputable medical doctor or scientist would ever believe that color could be effective as a therapy. I wrote a letter to the journal and never received a response. I pointed out in my letter that an entire volume from the Annals of the New York Academy of Sciences was dedicated to "The Medical and Biological Effects of Light." [3]

The use of color therapy does seem to be very simplistic for the major medical problems that we are confronted with today. The reason it seems so simplistic is the conditioning we receive from medical doctors saying that difficult medical problems require difficult medical solutions.

As I was proofing this chapter, in the background I heard a television commercial promoting products for arthritis pain. It was an advertisement warning users of several of the pharmaceuticals for arthritis pain that this class of pharmaceutical products can cause internal bleeding and ulcers that can lead to death. Death! The advertiser made the point that viewers should ask their doctors if the

---

[3] *The Medical and Biological Effects of Light*, Annals of the New York Academy of Sciences, Vol. 453, 1985.

risks are worth the benefits for their conditions. This is ludicrous! These medical therapies are worse than the original problem. How has our medical approach fallen to the point where we have to make health choices between one ailment and another as side effects of our therapies? We have gotten lost, and we must find our way back to more natural approaches that exist. There are answers to our problems, but we must search in a more open-minded manner.

The use of color dates to 500 BC, and there has never been a report of anyone who has died from using color therapy. Today there are more than 900 books written on the use of color therapy. Thousands of studies and experiments have been conducted. Most of the research is relegated to animal studies showing how color and light affect various functions in rats. I once contacted a researcher who had been doing studies on rats for his whole career, and I mentioned that color had been used on people in the 1920s and 1930s and was extremely effective, safe and well-documented. His answer was that this was what he was allowed to do, and he would not take a chance in jeopardizing his position to expand his research to humans.

I believe that the reason color therapy is not in more common use is that it is not a good moneymaker for the medical profession. A person can use simple color gels or slides and use any type of light source or even a slide projector. The cost of the correct color slides is less than $30. A color therapy session lasts about one hour and uses simple electricity to drive the light source, and because anyone can use color therapy at home, no revenue is produced for a medical doctor or clinic.

I have personally witnessed the effectiveness of color therapy in nearly five hundred people since 1977. I have seen color therapy work as effectively on animals as it does for people. In fact, because of a family emergency I had a personal experience with the effectiveness of color therapy. In 1984, my father was admitted to a leading hospital in Southern California to get an examination for back spasms. While he was in the hospital, he was given a mylogram, a dye injected by a needle, and apparently as a result of something on the needle, he ended up with an infection. Within hours of the test, every part of his body was shutting down, and he was rushed into surgery. The medical team spent four hours in surgery, but after a few days, the head of neurosurgery stated that my father was going to be a quadriplegic.

The surgeon told my mother that the hospital would begin quadriplegic training when my father was able to handle it.

My father remained in intensive care for more than 3 ½ weeks, and the doctors added that my father's speech would be permanently impaired.

After my father left intensive care, I spoke with the head of neurosurgery and told him we would be using color therapy on my father. The doctor asked if our therapy would disrupt any of their devices or mechanical or electrical machines in the hospital. I assured the doctor that our therapy was safe, quiet and effective. He did not have anything that would work for my father, but I believed that the color would work. What a strange situation when I thought back to my father wondering if I would ever discover the right color for various health conditions. I did have the right color for my father's condition.

My mother projected the color on my father for one hour in the morning and one hour in the evening. Interested hospital staff would walk by watching us without ever commenting about what we were doing. After a few days, my father began wiggling his toes. After another few days, my father could move his feet. One day the head surgeon, who hadn't seen my father for a week, was making his rounds and mentioned to my father he would like to see him try to wiggle his toes. The surgeon placed his face very close to my father's feet. My father told him he should step back or he might get kicked in the face. The surgeon laughed and thought my father was joking. When my father raised one foot completely off the bed the surgeon was in a state of shock.

My father eventually progressed to a wheel chair, then a walker, then crutches and finally was walking with two canes. Six weeks later, aided by two canes, my father walked out of the hospital.

If you recall in the early 1980s, the medical doctors made it difficult for you to see or get your own medical records. One of the hospital staff had made a copy of my father's medical chart, and I accidentally walked off with it. It states, "The patient has made a complete recovery from quadriplegia to 4/5 strength throughout his body…the family has refused further urological workups. He has a neurogenic bladder that will require a catheter."

Once my father returned home we used a special device called a multiple wave oscillator, and in two treatment sessions, my father was able to urinate on his own without a catheter. So much for the expert diagnosis of my father's medical condition!

I am very passionate about the use of color therapy, and I am sad about the treatment given to those people who have tried to make it available. Dinshah was a brilliant man who made an important discovery. To receive a compliment acknowledging him as being of the importance of Thomas Edison is high praise indeed. He was not a quack or someone looking to make fast money selling some sort of product. He was a serious researcher who spent his life in the service of people. Credentials, as we learned from our section about Dr. Nordenstrom, do not seem to be enough to allow someone to have a fair evaluation.

Can you imagine how difficult it would have been for Dr. Kate Baldwin in the 1930s to arrive at the position of Senior Surgeon at one of the oldest surgical hospitals in the United States? She felt so strongly about the use of color that she gave up her position rather than not be allowed to use her machines.

Pure greed keeps color therapy from being a part of our medical system. Even though I had the personal experience with my father, I cannot help but visualize the pictures of Grace Shirlow. What would Grace's life have been like without the use of color therapy?

Who is colorblind? Dinshah Ghadiali and many other color practitioners presented us with a natural and effective therapy - and nothing happened.

Some benefits of using color therapy:
- It is completely safe with no side effects.
- It is an inexpensive therapy that is easy to administer.
- It does not pollute the environment or have toxic ingredients.
- It can be utilized for many medical conditions.
- Color slides do not wear out or need to be replaced.

ONLY YOU CAN MAKE IT HAPPEN!

God gives every bird its food, but he does not throw it into the nest!

Chapter 5
Waves That Heal

Having a background in sports, then going into business and later spending twenty-five years studying the spiritual aspects of the mind/body connection, have provided me with a greater awareness of my own body, mind and spirit.

I believe that the longer you spend in study and self-introspection the less a division or separation exists between the body, mind and spirit. It is obvious to me that whatever activity or experience you have in one of these areas affects the other two to some degree.

Around 1978, I kept discovering references to an unusual therapeutic machine called a Multiple Wave Oscillator (MWO). What fascinated me about the machine was that it projected frequencies through the air and never touched the patient. The three or four references that I read mentioning the MWO also mentioned two books that described the theory and use of the machine. The books were *The Secret of Life* and *The Waves That Heal*. Although I was able to find the two books, one originally written in the 1940s and the other in the 1950s, it wasn't until 1983 that I had the opportunity to actually see the device itself.

In 1983, I met Phil and Allie who had several MWOs. Phil explained to me that the MWO was not only effective for alleviating physical health problems, but that he had found the machine of great value during meditation. Phil further explained that he had been able to place himself in deep meditation, and in this state was able to bring forth answers that were helpful in his business activities. Phil also knew several other people who had also used this technique. Here was a device that appeared to work on three levels: mind, body and spirit.

What struck me when I first saw the MWO was that it was the size of a brief case, and while it was operating it sounded like bacon sizzling on a hot skillet. Phil and Allie told me that they had been using the machine for several years and had been working with family and friends who had a variety of health challenges. When they told me some of the conditions, which included cancer and some other debilitating difficulties, I became very interested in learning more.

My meeting with Phil and Allie confirmed, in my mind, that the MWO was a device that might be effective and led me to reread the books that I'd found several years before.

George Lakhovsky (1869-1942)

Georges Lakhovsky invented the MWO and wrote a book of his discoveries titled *The Secret of Life*. [1]

Lakhovsky had great support from physicists and scientists, but only a few medical doctors seemed to be interested in his theories. One of Lakhovsky's strongest supporters was Professor D'Arsonval. D'Arsonval was a French physicist and physician. He worked under Claude Bernard and C. E. Brown-Séquard (whom he succeeded in 1897 at the Collège de France) and was a professor at The Sorbonne from 1894 to 1932. The D'Arsonval galvanometer is named for him. D'Arsonval was known as a pioneer in electrotherapy, and he studied the medical application of high-frequency currents and was involved in the industrial application of electricity.

The main premise of the Lakhovsky theory is that every living being emits radiations. Lakhovsky was working in France during the 1930s; and, other than D'Arsonval and a few colleagues, everyone else simply ignored his work. If Lakhovsky had been in the United States he would quite possibly have found support from several researchers including Dr. Harold Saxon Burr of Yale University, Dr. George Crile of the Cleveland Clinic, Royal Rife of San Diego and Dr. Otto Rahn of Cornell University. All of these researchers were working on theories that aligned with Lakhovsky's theories.

---

[1] Georges Lakhovsky, *The Secret of Life*, translated by Mark Clement, (Messrs. Heinemann Ltd., 1963)

In 1956, just a few years after Lakhovsky died, astronomer John Pfeiffer, nicknamed the "Boswell of Radio Astronomy," wrote a book titled *The Changing Universe*. Pfeiffer pointed out that every human being is an emitter of radio waves. We are all a living broadcasting station of exceedingly low power. Our stomach wall sends out not only infrared heat waves but also the entire spectrum of light - ultraviolet rays, X-rays, radio waves and so on. Of course, all these radiations are fantastically weak, and the radio waves are among the weakest. Pfeiffer stated that the fifty-foot aerial tower of the Naval Research Laboratory in Washington, the most accurately constructed aerial in existence, "...*could pick up radio signals coming from a person's stomach more than four miles away.*"

Our modern technology might be able to pick up even more subtle energies then those recorded in 1956. We should consider that we are more than just skin, bone and fluids. We have electromagnetic activities taking place within the cellular structure of our bodies.

Lakhovsky believed there are multiple, fundamental radiations that are the basis of all living beings. For Lakhovsky, each individual cell, the essential organic unit in all living beings, is nothing but an electromagnetic resonator capable of emitting and absorbing radiations of very high frequency. We remain healthy when our cells are transmitting and receiving their proper frequencies.

Lakhovsky believed the secret of a healthy life is when all cells are in their perfect balance and react with other cells in total equilibrium.

What is disease according to Lakhovsky? It is the oscillatory disequilibrium of cells, originating from external causes. It is, more specifically, the struggle between microbe radiation and cellular radiation. If microbe radiation is predominant, disease is the result, and when vital resistance is completely overcome death occurs. If cellular radiation gains the ascendant, restoration of health follows.

Lakhovsky believed that there are basic frequencies of energy for cells, tissues and organs that are present when the body is in a healthy state. If the energy frequency of the cell drops to a level that is lower than a microbe that is attacking it, then the microbe begins to take over.

Lakhovsky developed his MWO to broadcast multiple frequencies through the air to the cells located throughout the body and in different organs. Lakhovsky believed that the cells were like miniature broadcasting stations and receiving stations for frequencies.

Lakhovsky Multiple Wave Oscillator

Numerous researchers have validated Lakhovsky's idea regarding a cell's ability to broadcast and receive frequencies. These researchers have won the Nobel prize in physics.

Lakhovsky's idea was to send the cells of the body the frequencies they would normally be vibrating at in their healthy state. Cells that are diseased - that is, operating at improper frequencies - could then pick up their correct frequencies and align with them. Once the sick cells were restored to their correct frequencies and their healthy state, they were stronger than an offending microbe, bacteria, virus or other organism and could then overpower the invading microorganism.

I don't want to minimize the tremendous research that Lakhovsky must have conducted to discover what the correct frequencies were for various cells, tissues and organs. He began his research to develop a cure for cancer. He felt that the medical community was struggling in vain to find the answer to this terrible disease, and he wanted to help find a cure.

Lakhovsky started his experimentation with cancerous plants. Repeatedly he was able to demonstrate that his theory of electromagnetic waves was effective in a cure for cancer. In 1923, he

84

constructed his first MWO machine that created a constant electromagnetic field broadcasting the proper healthy frequencies to all the cells. Whether the cells were in a plant, animal or human being, this healthy field was used to bolster any weak cells in a system so that they could fight against any disturbing frequencies of an invading organism. Lakhovsky ran his MWO using alternating current that is very inexpensive - a plus for sure.

Lakhovsky was fortunate to draw the attention of several medical hospitals in the area where he lived. For the first six years of his research, he worked at the Salpetriere Hospital in Paris where he discovered the full field of frequencies that every healthy cell needed to operate at its natural efficiency.

Everything in the universe vibrates and oscillates, whether it is a microbe, human, plant, cell or rock. Lakhovsky moved from his plant experiments to using his MWO to treat patients with cancer at various hospitals in Paris, which included the Saint Louis, Val-de-Grace Calvary and the Necker hospitals.

Reports from those years showed that the patients he treated for cancer were still in complete and perfect health up to six years following his treatment.

A wonderful feature of the MWO treatment is that it does not attack any microbes or other invading microorganisms directly; it does not destroy any live tissue, but reinforces the vitality of the organism by accelerating the oscillation within the cell back to its natural and healthy condition. A healthy immune system defeats or resists any microbe or other pathogen that might be attacking it.

Unfortunately for Lakhovsky, when *The Secret of Life* was published in France, the Germans were just invading the country. Lakhovsky, who was Jewish, fled to the United States to avoid the German troops, and he never had the possibility of promoting his work to a greater audience.

In 1985, Phil introduced me to Ralph Bergstresser, of Phoenix, who was building MWOs. Ralph is one of the most amazing people I have met in my life. He worked as a consultant for the government and was very knowledgeable when it came to electromagnetic therapies. One point that Ralph was very proud of was that he was the last person to see Nikola Tesla before the great scientist died. One of the crucial working parts of the MWO is a Tesla coil. Tesla is one of the greatest inventors in history, if not the greatest. Ralph always kept me intrigued with the incredible contacts that he had made throughout his life. He shared with me that he had been present in offices within the government when it was decided to "place the lid" on all electromagnetic therapeutic devices and smother any new research. Ralph specifically mentioned several of the people and technologies discussed within this book.

Before Ralph died, he had sold several thousand of his briefcase MWOs and had followed the results of their use. During the last 20 years, I have personally witnessed more than **200** cases that the MWO has successfully assisted. I have personally witnessed cancer, multiple sclerosis, gangrene, ALS and many other debilitating problems be greatly assisted by the use of the MWO. Please note that the aforementioned conditions are not the same as curing a little headache or feeling a little better in the body. These are the worst medical conditions that we have in our society today.

While traveling, I have met many scientists and researchers in Europe who have built their own MWOs. **I am not recommending someone buy a MWO nor am I recommending any specific device; I am merely pointing out that this technology must be evaluated in an unbiased and open medical forum where it can be evaluated for its effectiveness.** During the 1930s and 1940s, none of the medical journals that reported on the cures derived by the MWO reported any adverse effects.

As I shared in chapter 4, my father was diagnosed with a neurogenic bladder and was told he would never urinate naturally without a catheter for the rest of his life. The Spectro-Chrome system had brought my father back from being a quadriplegic, but did not bring back the function of his bladder. Several weeks after my father had been home, his personal physician invited me over to his home to discuss the amazing results my father achieved. During our discussion, I asked my father's physician what spot on the body might assist the functioning of the bladder. He mentioned that a point near the base of the spine above the top of the crack of the buttocks was a collection of nerves, and he suggested I work at that site.

The next morning I used the MWO for ten minutes on my father. Several hours later, my mother called and mentioned that my father's pants were wet without using the catheter. The following day I stopped by my parents' home before returning to my home. I worked on my father for fifteen minutes and then left. By the time I reached my home, which was ten minutes away, my father was urinating on his own. He urinated on his own for the remaining eighteen years of his life.

The question is whether this was just a fluke or luck, or perhaps something that really helped. I believe that the answer is in the results.

Medical journals in France, Italy and other countries carried reports of the amazing results of the MWO on numerous medical conditions, all without any side effects. **Note this article published in the New York Times on Wednesday, April 4, 1928.**

Mark Clements translated *The Secret of Life* into English and wrote another book titled *The Waves That Heal.* In *The Waves That Heal,* Clement wrote about the amazing results Lakhovsky achieved in a large New York hospital where he practiced when he migrated to the United States. Sadly, Lakhovsky was not able to pursue his work. An automobile struck him while he walked to the hospital one morning, and he died from the accident. Ralph told me that he interviewed a nurse at the hospital who was looking out the window at the time of the accident. In her opinion, there was no question that it was deliberate. Hearsay for sure, but we'll never know.

was looking out the window at the time of the accident. In her opinion, there was no question that it was deliberate. Hearsay for sure, but we'll never know.

Lakhovsky brought a therapeutic device to the United States to help all people with their medical problems. He discovered a therapy that did not place people in a worse condition than their original problem did. He brought us something that he thought would help - and nothing happened.

YOU CAN MAKE IT HAPPEN!

Life is activation of inherent forces that are a part of your original creation.

Chapter 6
Fields of Life

During my years of research in the health field, I have read that nearly 98% of the human body is replaced every six months. Several researchers have stated that some of the organs of the body are completely renewed in even less time than that. The consensus is that our bodies do not remain stagnant but constantly change over the course of our lives. I have wondered how we keep our basic physical form if all of the cells are changing in our bodies, and at the same time this is happening, we have different aspects of metabolism taking place as we are digesting food and drink. What keeps our basic forms in the human shapes that we see them? For that matter, what keeps any living organism organized? When we eat food that is part of a plant or fruit, how do our forms keep their shapes?

When it comes to our bodies, I have also wondered if there is not some type of early warning system that would allow us to have foreknowledge that some health problem, illness or disease is about to appear. The warning would come before x-rays or cat scans or MRIs. By the time we see those images we already have the problem.

I would receive the answer to both of these points when I read the work of Dr. Harold Saxton Burr, Ph.D.

Dr. Burr's basic premise was that the universe in which we find ourselves and from which we cannot be separated is a place of Law and Order. This follows the Greek concept of *The Good* that we mentioned in a previous chapter. Events are not random accidents but follow discoverable causes. We might say that there are no effects that do not have causes, and miraculous events are merely the results of laws that are not yet recognized. I am not saying that there are not dramatic events that we witness when it comes to healing. We call them miracles when they happen. However, I believe that they take place based on specific laws that we will discover.

In an upcoming chapter, you will learn how two nuclear medical doctors made visible the subtle energy of the ancient Chinese acupuncture meridian system. Writings discussing this system date back more than three thousand years. It took the development of modern measuring equipment to scientifically validate the system's existence. Dr. Burr worked to develop special equipment to measure the organizing fields that he knew already existed.

Dr. Burr found that all matter is organized and maintained by an *electro-dynamic field*, invisible to our eyes, that is capable of determining the position and movement of all charged particles. He spent more than fifty-years experimenting with his theory and never met with any contradiction.

Dr. Harold Saxton Burr, Ph.D., was E. K. Hunt Professor Emeritus, Anatomy, at Yale University School of Medicine. He was a member of the

Harold Saxton Burr, Ph.D.
E.K. Hunt Professor Emeritus,
Yale Univ. School of Medicine

faculty of medicine for more than forty-three years. From 1916 to the late 1950s, he published, either alone or with others, more than ninety-three scientific papers.

Dr. Burr discovered that all living things, which include animals, plants, humans, microbes and even seeds, receive their visible forms, and are controlled by what he called *electro-dynamic fields*. He measured these *electro-dynamic field*s using standard voltmeters.

These *fields of life*, or *L-fields*, are the basic blueprints of all life on this planet. Their discovery is of immense significance to all of us. Dr. Burr believed that, because measurements of *L-field* voltages can reveal physical and mental conditions, doctors should be able to use them to diagnose illness before symptoms develop. If we were able to address conditions before they would normally develop, we would have a better chance of successful treatment.

I think this bears repeating. Before any physical symptoms are present or able to be pictured or measured on our standard diagnostic tests, the condition, disease or illness, produces a change in an *invisible* electro-magnetic field that surrounds our body and gives us the indication that there is going to be a change in our physical body. This allows us to make physical changes before the health problem manifests in a manner that would necessitate drastic or dramatic intervention.

Dr. Burr used an interesting analogy to help show what *L-fields* do and why they are so important. We know that when iron filings are scattered on a card held over a magnet, they will arrange themselves in the pattern of the lines of force of the magnet's field. In addition, if the iron filings are discarded and fresh ones scattered on the card, the new filings will assume the same pattern as the old. Something like this happens in the human body. Its molecules and cells are constantly being torn apart and rebuilt with fresh material from the food we eat. However, thanks to the controlling *L-fields*, the new molecules and cells are built as before and arrange themselves in the same pattern as the old ones.

Dr. Burr and his *L-fields* give the answer to the two central questions that I posed at the beginning of this chapter - how do our bodies keep in shape through the numerous activities of metabolism and how do we hold our shapes with constant regeneration of cells taking place. The concept of *electro-dynamic fields* of the body serves as a matrix or mold, which preserves the shape or arrangement of any material poured into it, however often the material may be changed.

When we look at a mold used in a kitchen or a factory, we know the final shape that the substance we place within it will turn out to look like. This is an analogy of what Dr. Burr knew when he examined the *L-field* of various objects with his instruments. When Burr examined the *L-field* in a frog's egg, he knew exactly the location of the frog's nervous system while it was still in its egg. The frog's *L-field* was the matrix that would determine the frog's ultimate form when it broke through its egg.

If we found a dent, nick or scratch in a mold, we would expect to find some dents or nicks and scratches in our final product. When Dr. Burr's measurements found an irregular *L-field* to what he was expecting, he labeled it as a battered *L-field*. It was battered because it had abnormal voltage patterns, and it gave a warning of something out of shape in the body, sometimes in advance of actual symptoms.

For example, malignancy in the ovary has been revealed by *L-field* measurements before any clinical sign could be observed. Such measurements, therefore, could help doctors detect cancer early when there is a better chance of treating it successfully.

Dr. Burr was adamant when he declared that it is impossible to avoid the conclusion that living mechanisms possess *electro-dynamic fields*. Dr. Burr, as most of the other scientists whom I have written about in this book, knew from their experimentation that the body is an electrical and magnetic device. Burr was able to prove that *fields of life* are present for every object that exists, and he had the ability to measure that field.

It appears that during all growth and development that takes place in our bodies, there are significant variations in the electrical patterns that can be measured in the same manner as heart and brain waves. The changes in electrical activity can be measured during activites in the generative tract associated with ovulation and during the development and progression of cancer.

An associate of Dr. Burr, Dr. Louis Langman, M.D., F.A.C.S., of New York University and Bellevue Hospital Gynecological Service, decided to experiment with the diagnostic aspect of the *L-field*. Dr. Langman examined more than 1000 female patients within his hospital who had a variety of health problems including fibromas and other problems in their generative tracts. Then he narrowed his research to those patients who showed a marked change in the voltage gradient between the cervix and the ventral abdominal wall.

There were a hundred and two cases where there were significant shifts in the voltage gradients, suggesting a malignancies within the patients. Surgical confirmation was found in ninety-five of the one hundred and two cases. Doctors were able to identify successfully the existence of a malignancies in the generative tracts through biopsies. Malignancies were located within a wide area of the generative tracts, and were not limited to only one spot.

Dr. Burr decided to examine the electrical properties of cancer-susceptible mice to determine if the voltage measurements would change during the initiation and growth of cancer tissue. Dr. Burr decided to implant cancer tumors and then see if there were changes in the voltage measurements. Remember that Burr was doing this in the 1930s. Burr stated that the results of the experiment were surprisingly consistent. Twenty-four to twenty-eight hours after the implantation, changes were observed in the voltage gradients. This differential increased steadily and quite smoothly to reach a maximum of approximately five millivolts on or about the eleventh day. In the slow-growing tumors, potential differences began to emerge on the third or fourth day but reached their maximum of approximately three millivolts on the tenth or eleventh day.

What does this all mean? It means that as a cancer tumor is growing there is a change in the electrical potential at the site of the tumor. There is a measurable electrical difference between the cancer tumor and the normal tissue surrounding the tumor. This was exactly what Dr. Nordenstrom discovered with his research in the 1980s.

Difference in voltage between tumor and surrounding tissue.
Discover Magazine, April, 1986.

In the normal animals, there was no significant fluctuation in voltage. Healthy animals did not have a change because cancerous material was not implanted into them. What is exciting about this research is that it was not just a subjective opinion that something is different with an animal with cancer, but there are measurable and reproducible electromagnetic values.

Dr. Burr's experimentation proved that there are matrix fields that establish the mold of the entire object, but he was also convinced that there must be even smaller subsidiary or local fields that surrounded the component parts of a living organism. He knew that there is the field that determines the general form, but he also discovered that there are fields in ovulation and malignancy.

Dr. Burr soon discovered *electro-dynamic fields* becoming active when a subject had sustained an injury. Definite electrical changes were taking place during nerve injury, and the changes in the measurements were not the result of vascular or sweating responses. Dr. Burr was able to measure the changes taking place in the *L-field* during the healing process. During all phases of the curative process, gradient measurements show the body initiating various functions pertaining to wound healing.

Every biological and functional activity within the body had a corresponding relationship to electric or voltage changes taking place in the body. This is exactly the discovery made by Dr. Nordenstrom.

Dr. Burr was not content with only the physical aspect of our reality because he also decided to investigate if electro-metric technique might be useful in the neurological and psychiatric fields. He conducted numerous experiments with Dr. Leonard J. Ravitz, Jr., at one time on the staff of the Department of Psychiatry at Yale.

Dr. Ravitz found that he could establish baseline voltage gradient measurements for individuals who had normal mental functioning. He also discovered that different psychological problems had their unique gradient measurements. With this information, Dr. Ravitz could determine if particular therapies were helping a patient as well as providing an additional tool as to when a patient could be discharged from the hospital.

Dr. Ravitz, who worked with Dr. Milton Erickson, did numerous experiments with patients under hypnosis as well as patients who were experiencing emotional disturbances such as grief or shock. Dr. Ravitz conducted an experiment with a patient under hypnosis who was told to relive an experience of grief from the patient's past. As the patient was mentally going through the experience, the doctors could measure specific changes in millivolt measurements. Dr. Ravitz found that an investigator using the measuring devices did not need to know the mental state of patients beforehand, but his findings tallied closely with the psychiatric diagnoses done by a professional.

One of Dr. Burr's most exciting discoveries dealt with the recording of voltage changes during female menstrual cycles. His research began using female rabbits because female rabbits ovulate about nine hours after stimulation of the cervix. The researchers stimulated a female rabbit, connected electrodes, and then watched through a microscope to see if there were any electric changes as the rabbit's body was going through its physical activity. The researchers saw that when the follicle ruptured and the egg was released, there was a sharp change in the voltage on their recorder.

We know this technique is used today to assist women in determining the best time to become pregnant. It was Dr. Burr and his associates who did the research leading to this discovery.

Dr. Burr would not have made his great discoveries if there had not been a change in the technology of electromagnetic measuring

93

devices during his life. The older techniques of electrical measurement were complicated by the difficulty of separating voltage, current and resistance. When the radio vacuum tube was developed, this changed the entire field of measuring equipment.

In 1935, with several of his collaborators, Dr. Burr developed a measuring technique called The Burr-Lane-Nims technique.

Perhaps more than any other researcher in the field of electromagnetics, Dr. Burr could make the bold statement that wherever there is life there is electricity. Electrical phenomena are present in every aspect of our existence and the functional activities that take place within our bodies.

Dr. Burr discovered that voltage differences are not chaotic but organized into characteristic patterns correlated with species, with age, and, in all probability, with the individual.

However, the most important part of this research is that Dr. Burr and his associates were able to measure changes in electrical patterns weeks before cancer tumors became obvious in mice. In their observations of genetically controlled strains of mice, they found that spontaneous adenocarcinoma of the mammary gland could be recognized by a change in the electrical pattern of potential differences weeks before it was evident by touch. They also discovered that some mice had very strong electrical patterns and stayed resistant to the cancer. There was a significant difference in electrical patterns between cancer-susceptible mice and cancer-resistant mice.

Dr. Burr stated, "It is impossible to avoid the conclusion that living mechanisms possess electrodynamic fields, and that they are stable, being modified only by profound changes in biological activity. Growth and development produce just as significant variations in the electrical pattern as do heart and brain waves. The activity of the generative tract associated with ovulation, the development of cancer, and in all probability many yet undiscovered functions of the living organism, are controlled and regulated by electrodynamic fields. It may well be, therefore, that here lies the long-sought clue to the problem of organization, disturbances of which results, among other things, in the wild, unrestrained atypical growth of cancer." The major discoveries of Dr. Burr can be found in his book, *Blueprint for Immortality*.

Dr. Burr pointed the way - and nothing happened.

YOU CAN MAKE IT HAPPEN!

94

"To act is easy, to think is hard." – Goethe

Chapter 7
The Phenomenon of Life

The story of the Buddha relates that he was born into a wealthy family and had everything that he could want. His father, King Suddhodana, wishing for his son to be a great king, shielded him from knowledge of human suffering. At the age of 29, Buddha left his palace in order to meet his subjects. Despite his father's effort to remove the sick, aged and suffering from his view, Buddha was said to have seen an old man. Disturbed by this sight, and when told that all people would eventually grow old, the prince went on further trips where he encountered a diseased man, a decaying corpse and an ascetic. Deeply depressed by these sights, the Buddha sought to overcome old age, illness and death by living the life of an ascetic.

Unexpected experiences in life can lead to wonderful discoveries along with establishing the future direction of a person's life. I've been blessed with several profound and unexpected experiences that placed me on a new pathway during my life. Some of my experiences came during interpretation of nightly dreams, and other experiences came in the midst of an investigative journey. Both of these experiences have led to a deeper introspection and a change in the course of my life.

The same is true for George Washington Crile. Dr. Crile is a famous United States surgeon who was one of the original founders of the Cleveland Clinic in Ohio. Dr. Crile is formally recognized as the first surgeon to have succeeded in performing a direct blood transfusion. He contributed to many other procedures including describing a technique for using opioids, regional and general anesthesia which is known as balanced anesthesia, and procedures for neck dissection.

Further tribute to Dr. Crile includes the SS George Crile, a United States liberty ship, named after him during World War II, and a lunar crater named after him. At the time of this writing, another award has been given to Dr. Crile. On January 16, 2008, The Capitol Square Foundation chose six individuals to receive the 2008 Great People of

Ohio award. The other recipients include Jesse Owens, Thomas Edison and Harriet Beecher Stowe. [1]

Dr. Crile's life work began because of a life-changing experience that he had in 1887. During his first day as an intern at University Hospital, Cleveland, a young man whose legs were crushed under the wheels of a streetcar was brought to the hospital. Crile recognized the man as a fellow medical student whom he had met during his studies. The man's name was William Lyndman, and he was in a state of profound shock.

Although it was a severe accident, there was little loss of blood. However, Lyndman's legs were damaged to the point that the senior surgeons decided that they needed to amputate his legs at the thigh. After the operation, Crile watched over his friend throughout the night. During the night, Crile noted Lyndman's steadily failing faculties and deepening depression. Dr. George Crile witnessed his first death when Lyndman died the next morning.

George W. Crile

Crile completed a postmortem after the death, but noted that all of the organs and tissues appeared to be normal in their physical structure. If the organs were intact and everything seemed in working condition, then he wondered why his friend died. According to Crile's medical studies, Lyndman's body should still be operating, and his friend should still be alive. The death would make more sense to him if there were a substantial loss of blood or one of the organs was severely injured. However, none of these factors was present.

While sitting with his friend throughout the night, Crile noticed that life seemed to drain slowly out of Lyndman as one organ after another began to fail. This also confused Crile. Why would the organs fail if blood was still circulating through the system and none of the organs was damaged. This observation led Crile to realize that there had to be more to the human body than what he was learning through his medical textbooks. Crile began to believe there had to be some type of living energy or vital energy that was an aspect of what he had witnessed with his friend. This idea initiated a quest that would

---

[1] Ohio Statehouse: Communications: Press Releases, Colombus, Ohio, January 16, 2008.

carry him throughout the rest of his life. He was seeking the answers to what provided life for his friend William Lyndman - and what caused his death.

I am not going to go into the massive amount of research that Dr. Crile compiled during his life. Many times Crile postulated theories only to discover other factors that he had not considered. Each new discovery led him to formulate even greater conceptions of what the driving force might be in the human body.

In order to discover what the phenomenon of life might be, Crile began his quest looking at the reason why people die from shock when there is no apparent physical explanation. In cases of shock, he found no apparent change in circulation, in respiration or in the blood. Death did not appear due to failure of the kidneys, the stomach and intestines, the pancreas, the spleen, the thyroid gland, the muscles, the tendons, the connective tissue, the bones or the joints. Therefore, Dr. Crile needed to look deeper within the make-up of the body.

Crile spent ten years experimenting on more than 2,500 test animals and then examined every cell of all the tissues and organs of their bodies. Through microscopic analysis, he discovered that at death there were changes in particular cells of the thyroid, liver and the adrenal glands. He found that the most significant changes appeared in the cortex where the cells appeared to have diminished in size.

Because of his continued examination, Crile realized that the thyroid, liver and adrenal glands were the Rosetta stone for his discovery of a controlling energy that governs life. When the subject is alive these cells operated in a particular manner, and when death occurred particular cells withered and died despite the presence of blood and normal circulation.

Crile reasoned that if all of the fluids in the system are intact at death, then he must be dealing with an energetic situation taking place within the body. Science is aware of the fact that a cell contains a comparatively acid or positive nucleus and an alkaline or negative cytoplasm and gives the strong appearance that living cells are electric in nature. This led Crile to formulate the electrical or Bipolar Theory of the living process. He also termed the presence of electricity within the cellular structure as the Radio-Electric Theory.

Crile believed that he was on the right track in his quest, and realized that he would not find his answer in physiology, biochemistry or morphology but in the area of biophysics. All of his research was now involved in measuring changes in temperature, in electric conductivity and the electric capacity and electric potential of cells.

Crile subjected his test animals to injury, emotion, infection, stimulants, narcotics, insomnia, adrenalin and other factors while measuring what took place in the cells of the thyroid, adrenal glands and liver from an electrical standpoint. In all the test cases, Crile discovered an exact corresponding change in electrical activity within the cells.

Crile decided to carry his research into the examination of cells in the structures in the plant kingdom. He studied fruits and vegetables, which he subjected to loss of oxygen (asphyxia) and to anesthetics and stimulants. Crile examined cells for changes in temperature, electric conductivity, electric capacity and electric potential. What he found was that the changes in these measurements corresponded closely to what was taking place with the cells in the three organs of the animals.

Protoplasm is the essential matter that makes up all plant and animal cells, and Crile's research proved that the protoplasm is a system of generators, conductance lines, insulators and an infinite number of thin films that hold electric charges.

The next leap for Dr. Crile was his realization that all matter within the universe is affected by the same energy and that all matter is electromagnetic in its nature. Electromagnetic effects sustain nature, plants, animals, humans and all living things. The great source of radiant energy is the sun. The sun provides energy for plants and the nitrates that feed the plants. Animals eat the plants that provide radiant energy for animals and humans when they eat either plants or animals.

Crile believed that the animal kingdom is highly sensitized to radiant and electric energy, and this energy is what gives life to the body. If there is a change in the body due to exhaustion, injury, emotion or infection, this leads to a change in the stored energy that governs the rate of internal combustion. The chief organs of the body that process this solar or radiant energy are the brain, thyroid, and the adrenal-sympathetic system. Crile's research showed that the thyroid gland and the adrenal-sympathetic system energize the body by shifting the emitted radiation toward the shortwave field.

Crile further conducted research within the electric fields of cancer cells and cancer tumors. In addition, he examined the effect of anesthetics and narcotics on cellular structure. Crile discovered that when there are changes in the electric potential of the cell, then the process of oxidation is interrupted and there is a corresponding change in the cell that can lead to mutation or death. The mutated cells would lead to a tumor that would continue to exist until the electrical balance was brought back to a normal condition.

The validation that Crile had come close to the answer that he sought was that at the time of death almost all electric activity ceased within particular cells of the organ even though all physical and chemical structures were still in place. This is what took the life of his friend Lyndman.

Dr. Crile had an elaborate but perfectly logical theory of how the human body operates on an energetic level. Protoplasm is constructed in a perfect manner to capture, adapt, transform and circulate the radiant energy throughout the system to bring about life. Crile found little particles existing within the body, which he called radiogens. These minute particles control the entire electromagnetic aspect of the life force that courses throughout the body.

Crile believed that just as the sun is the major generator of the energy that radiates to earth and provides life to everything that exists on our planet, there are billions of tiny suns within the nucleus of each cell that radiate light or energy within the body. All of Crile's research work goes hand in hand with the research of Dr. Nordenstrom, Lakhovsky and Dr. Burr.

Crile's Radio-Electric Theory identifies the mechanisms of how all our bodily functions are controlled and regulated. This theory leads us to a greater understanding and explanation for the mechanisms of memory, reason, imagination and various emotions.

Crile also found that during the healing process taking place in the body there were electric potential changes within the cells and tissues. He found that even the smallest microorganism had its own electrical activity that was a part of its life process. All life consists of cells, and all organisms are bipolar mechanisms constructed and energized by radiant and electric energy.

Dr. Crile did not build any therapeutic device nor did he announce a cure for cancer or other diseases. Crile gives us all hope for a better understanding of who we are and what we need to do to protect our vital energy. We must look at our energetic system as the basis of our lives and as the vehicle to keep us healthy.

He tried to tell us - and nothing happened.

"An acre of performance is worth the whole world of promise." - Howell

Chapter 8
Twice Normal Speed

In 1976 I was watching the Olympic games being broadcast on ABC's Wide World of Sports. On one of the segments, sportscaster Jim McKay was commenting on how amazing it was that Lasse Viren was running and leading the 10,000 meter finals. This came about because Lasse Viren had earlier won the 5,000 meter finals, and the concern was that he might not be ready to run the 10,000 because of a hamstring injury he sustained in the earlier race.

The television show kept flashing back and forth from Viren actually running the 10,000 meters to the trainer's room where earlier he had used an incredible machine that was allowing him to even be competing. This amazing machine was said to accelerate the healing process.

The Olympics allowed me to learn about this incredible device that had been documented by dozens of United States and foreign universities to accelerate wound healing at almost twice normal speed. At the same time, I would receive an important education of why we don't always learn about these great innovations. I mention this because the first bit of information I learned was that the FDA had banned the Diapulse as a quack device in 1972. In 1987, a judge would rescind that order and force the FDA to announce the Diapulse machine as a new emerging technology in medicine. [1]

To better explain this machine, I am including, in full, a report that was given to me by Robert Maver, FSA, MAAA, Director of Research for Mutual Benefit Life Insurance Company. In my capacity as a consultant for Mutual Benefit Life I introduced Robert to the Diapulse machine, and he made this report for his company and the insurance industry:

"A remarkable machine that uses pulsed electromagnetic frequencies to accelerate healing appears poised to make a dramatic impact on medicine and medical costs. Diapulse has been described to truly revolutionize much of the practice of medicine."

---

[1] Emerging Electromagnetic Medical Technology, May 25-28, 1989, Sponsored by The University of Tulsa Department of Continuing Education Office of Research and FDA Center for Devices and Radiological Health

"The Diapulse machine was developed in 1932 by Abraham J. Ginsberg, M.D., and Arthur Milinowski, a physicist. Ginsberg received medals from the U.S. Government for his invention of the 'sniper-scope' used on the M1 rifle. His good friend **Albert Einstein** worked out some of the mathematics for him in the application of electromagnetism to medicine. The first paper on Diapulse was published in 1934[2], and in 1940, animal studies were performed at Colombia University to prove the efficacy and safety of Diapulse.[3] In the 1950's, the Director of the Tri-State Research Program for the U.S. Government concluded Diapulse was safe and effective.[4]

Diapulse in use during Olympic Games

"In the late 1950's, Diapulse established research in the United States and at international universities and hospitals. There now exists a vast amount of research including laboratory, animal and control blind and double-blind clinical studies on the acceleration of wound healing with Diapulse therapy spanning more than 40 years. Because

---

[2] A.J. Ginsberg, "Ultra-short radio waves as a therapeutic agent, Medical Record, Dec. 1-8, 1934
[3] H.R. Halsey, "Histological Examination of Rabbit tissue", Personal Communication, Colombia   University, College of Pharmacy, Dec. 1940.
  H.R. Halsey, "Physiological Responses to Diapulse Therapy in Rabbits and Rats", Personal Communication, Colombia University, College of Pharmacy, Oct. 1941.
[4] Jesse Ross, "Results, Theories, and Concepts Concerning the Beneficial Effects of Pulsed High Peak Power Electromagnetic Energy" (Diapulse Therapy), Bioelectromagnetics Society 3rd Annual Conference, Washington, DC, August 9-12, 1981

the Diapulse effectively addresses three basic processes involved in healing, i.e., elimination of edema, absorption of hematoma and increased blood-flow, it has applications in a wide variety of medical conditions."

"In 1958, the Mayo Clinic confirmed studies previously carried out by Ginsberg demonstrating the cellular effects of pulsed electromagnetic energy known as Pearl-Chain phenomena. This study was repeated again at the New England Institute for Medical Research in 1961." [5]

"A 1961 Study on Arthritis produced some dramatic results.[6] The data reported on 63 cases of longstanding duration (average = 12 years) that had failed a variety of modalities. An average length of Diapulse therapy was 19 treatments over a period of six weeks. 59 out of 63 patients showed marked improvement. The study uses categories not usually seen, e.g., the table labeled 'Physical Aids No Longer Required' is reproduced below:

Crutches Discarded	5
Wheelchairs Discarded	4
Cane Discarded	1"

"A 1962 study of Diapulse in the treatment of pelvic inflammatory disease reported an average hospital stay of 7.4 days for Diapulse treated patients versus 13.5 days in the control group." [7]

"A 1964 double blind study on wound healing in surgical patients demonstrated an average shortening of hospitalization by 1.5 days in the Diapulse treated group." [8]

"A 1966 placebo controlled study of tonsillectomies demonstrated a 50% reduction in hospital stay for the Diapulse treated group. This was highly statistically significant."

"A 1969 report from Tel Aviv University Department of Plastic Surgery reviewed cases of soldiers requiring reconstructive surgery at

[5] A. Wildervanck, K.G. Wakim, et al., "Certain experimental observation on a pulsed diathermy machine," *Archives of Physical Medicine and Rehabilitation*, Vol. 40, Feb. 1959.

[6] Dr. Euclid-Smith, "Report on 63 Case Histories, Private Communication, 1961

[7] M. J. Lobell (Harlem Hospital, NY) "Pulsed high frequency electromagnetic radiations and routine hospital antibiotic therapy in the management of pelvic inflammatory disease. A preliminary report," *Clinical Medicine*, 69 (8), Aug. 1962.

[8] B. M. Cameron, (St. Luke's Hospital, Houston, TX) "A three phase evaluation of pulsed high frequency, radio short waves", (Diapulse on 646 patients), Am J. Orthopedics, March, 1964.

the end of the six-day war in June 1967. It was reported that through the use of Diapulse hospitalizations were reduced 25%."

(I might add here that all of these took place before the machine was banned as a quack device—Author)

"A randomized trial was conducted in England in 1983 on various hand injuries. It was reported that on the 7th day, only one patient in the treated group had slight loss of function still remaining; the other 29 had been discharged. This compares with only 3 patients in the control group who had been discharged in 7 days and the other 27, who still had symptoms of swelling, disability and pain on the seventh day."

"A study of Diapulse on spinal cord injury conducted in Poland in the late 1970's produced amazing results. 97 patients underwent treatment. A pronounced neurological recovery was observed in 38 patients, i.e., some 40% of the group. Remarkably, in 28 individuals the recovery had substantial functional value: the patients were discharged from the Neuro-orthopedic Department with paresis slightly impairing the function of the extremities. In other non Diapulse treated patients with such neurological lesions observed at the time of administration, one rarely attains a definite neurological improvement." [9]

"Further work was conducted at New York University Medical Center in the early 1980's on cats with induced spinal cord crush injuries. These double blind studies produced some phenomenal results. In a letter summarizing his work, Dr. Young, Director of Neurosurgical Research Labs at NYU, reported that 'the eventual percentage of cats receiving Pulsed Electromagnetic Frequencies (PEMF) at 4 hours after injury that recovered walking at 4 months after spinal cord injury is 78% (7/9), compared with 0% (0/10) of the sham treated injured controls. No pharmacological treatment that we have tested to date, including naloxone and methylprednisolone (both of which are in double blind randomized clinical trial in 12 spinal

---

[9] Jerzy Kiwerski, Teresa Chrostowska, "Clinical Trials of the Application of Pulsating Electromagnetic Energy in the Treatment of Spinal Cord Lesions, Chir. Narz. Ortop, Poland, Vol. 45, no. 3, ,pp. 273-277.

injury centers around the country) has exhibited this degree of effectiveness in similar animal studies.'" [10]

"Research recently completed at another hospital in New York followed up on a 1978 report from Poland on healing of decubitus ulcers (bedsores) with Diapulse as adjunctive therapy. In the New York study, 22 out of 22 ulcers were completely healed. Many had been deteriorating for months, some for more than a year. One ulcer, unhealed for 52 weeks with conventional treatment was healed in 1 week with Diapulse. Another unhealed for 168 weeks was healed in 7 weeks. The study will be reported in medical journals in 1990 with emphasis on the savings in hospital stays for this heretofore un-solved major healthcare problem." [11]

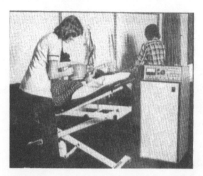

Diapulse in use in a London hospital

"Accelerated healing of burn victims offers an area of great promise. In a report on some 2,000 patients covering 9 years, results were described as significantly positive, especially in the almost immediate relief of pain. One attempt to confirm these observations with the rigor of double blind study was aborted when nurses involved in the study refused to participate after 3 days. **It was so apparent which patients were receiving the placebo machine.**"

"Clinical studies of Diapulse in the treatment of head injuries have also been conducted. 100 patients with similar Glasgow Coma Scale (GSC) sores of 8 or less were randomly assigned to Diapulse treatment or a control group. Untreated patients showed little or no

---

[10] Wise Young, (Department of Neurosurgery, New York University), "Pulsed Electromagnetic Fields Alter Calcium in Spinal Cord Injury", presented at The Society of Neurological Surgeons, 75th Meeting, New York, NY, April 25-28, 1978.
[11] This study did appear in *Decubitus*, February 1991.

improvement on the GCS, whereas the Diapulse treated patients showed significant improvement on the Scale."

"Sports medicine is another area that will be greatly impacted by Diapulse therapy. Just prior to the 1968 Mexico Olympic Games one of the European teams alerted the Olympic Organizing Committee it would be bringing a Diapulse machine to the Games. The Organizing Committee's initial reaction was to not allow the machine because it would give that country's teams an unfair advantage, especially in a sport like boxing. The Committee's final position was to request Diapulse Corporation to make available 30 machines at the Games for use by all teams. Other than oxygen it was the only therapy supplied by the Olympic Organizing Committee. They have been treating athletes at every Olympic Games since. In the Montreal games, multiple gold medal winner Lasse Viren of Finland gave credit to the Diapulse for allowing him to compete and win one day after a hamstring pull. Princess Anne used it after taking a bad fall in the equestrian and went on to win a medal in that event."

"Diapulse should also play a major role in occupational medicine. In 1964, R.G. Young, M.D., Medical Director Fisher Body Division, General Motors, Marion, Indiana, presented a paper on use of Diapulse at his plant. (*I remind the reader that this was before the ban—* Author.) More than 2,500 patients were treated in a 20-month period with beneficial results. Of particular note is the excellent result obtained on an assembly line doing repetitive motion work and subject to carpal tunnel syndrome. Dr. W.H. Caney, Jr., Watertown, Connecticut reported excellent healing results with Diapulse after trauma in industrial accidents. He characterized Diapulse as mandatory equipment for every industrial medical department."

"Today (1990) Dr. Tarasenko, Director of Medical Services in Mutual Benefit Life's Newark Home Office, is using Diapulse to treat a variety of ailments among employees. (*This was due to the fact of the recommendation I made to Robert and Dr. Tarasenko to investigate this device-*Author.) His experience has been so positive that he is preparing a paper on 40 cases to be presented at the national meeting for occupational medicine. One of his more celebrated cases is that of a senior executive for a competitor insurance company. This officer was disabled, working only 2 hours per week and unable to take a business trip because of excruciating back pain. He had a double laminectomy, nerve block, etc. etc. all with no relief. In two weeks treatment with Diapulse, he is back at work full time, taking business trips and out of pain. His company's Medical Director has recently purchased a Diapulse machine for its employees."

"Plagued by the 'tomato effect' science has finally caught up with the Diapulse and it is fully approved by the FDA In fact, Springer-Verlag has recently published <u>Emerging</u> <u>Electromagnetic</u> <u>Medicine</u>, edited by O'Conner, Bentall and Monahan. *(They would not have approved it without the court stepping in and ruling against the FDA-*Author.)

"The United States Military has shown high interest in the Diapulse as a result of a study recently conducted at Brooke Army Medical Center. This double blind, randomized prospective trial was conducted on acutely sprained ankles, the leading cause of disability with respect to lost training days. Noting significant reduction of edema following one treatment with Diapulse, Army researchers, presenting a paper at the December, 1989, annual meeting of the Society of Military Orthopedic Surgeons concluded: 'Application of this modality for reducing edema in acutely sprained ankles or other traumatic injuries could significantly decrease time lost from military training and result in large fiscal savings.'"

"The prospect of a new treatment modality that is safe, effective and inexpensive should be welcomed by our industry." [12]

I would like to mention an important point at this time. The Diapulse Corporation was not aware that I was writing this book and I did not consult with them regarding this chapter. The Diapulse Corporation prefers not to discuss the odyssey of their 15 year battle with the FDA. I get very disturbed when I think of how many people lost the benefit of this therapy while it was banned in the United States.

The FDA only allows the Diapulse Corporation to speak about the effect on pain or swelling. But what about all of the other aspects that you have read about in this chapter? I am not necessarily endorsing the Diapulse and I absolutely have no vested interest nor do I own stock in the company. But, how could such devices not be major parts of every hospital, clinic and treatment center?

The Diapulse presented its effective technology for 15 years - and nothing happened.

---

[12] Robert Maver, FSA, MAAA, Fall, 1990.

Chapter 9
Bypassing the Bypass

During a 1987 trip to Germany I had the opportunity to investigate an intriguing story relating to how a special herb from Africa had the ability to eliminate the need for heart bypass therapy in almost 90% of cases. During my visit I had a chance to meet with Dr. Kern who had discovered the value of this herb and had been researching it throughout his life.

In 1994, the Stuttgart offices of the World Research Foundation was given the complete collection of materials from the Gesellschaft fur Infarktverhutung (The Association for Prevention of Heart Attacks). The organization was founded on September 12, 1968, for the purpose of gathering and dispensing information for the prevention and treatment of heart attack. More than likely it is the best collection of books, articles, reports, studies and documentation dealing with the heart and heart glycosides anywhere in Europe.

In addition to gathering information, the Association arranged studies at several universities to investigate innovative treatments and to determine the effectiveness of several new preventive techniques.

One of the most interesting projects of the association was its research on the heart-glucoside G-Strophanthin.

Three thousand medical doctors in Germany submitted patient information concerning this plant therapy and its effect on angina pectoris (heart attack) demonstrating its effectiveness in preventing the need for heart bypass therapy.

I have written the following brief overview of the work of Dr. Kern in collaboration with Robert Maver and Prof. Dr. Karl Walter. This is not intended to be a medical article, and I am keeping it as simple as possible. This article was originally written in 1989 from the aforementioned information that at that time was held by the association of medical doctors.

As a result of the gathered information regarding G-Strophanthin, in 1989, Mutual Benefit Life submitted the information to the University of Colorado at Denver, Center for Health Ethics and Policy Graduate School of Public Affairs. This particular group evaluates health technologies and gives its opinion on the quality of the information and the likelihood that it will be received into the medical community. During the period I consulted for Mutual Benefit Life, I was given the report, a part of which is included at the end of this chapter.

Coronary artery disease is currently the leading cause of death in the United States. Despite the increasing sophistication of surgical techniques and a number of new drugs (e.g. beta blockers, calcium antagonists), it is estimated that more than 1 million heart attacks will occur this year resulting in 500,000 deaths. In short, we do not have an adequate therapeutic solution to the problem of myocardial infarction (heart attack).

The cornerstone of therapy for treatment and prevention of myocardial infarction is to remove blockages in coronary arteries that are thought to be the cause of the infarction. This adheres to the widely accepted coronary artery thrombosis theory of infarction, that arteries become clogged with plaque, damaged from such things as smoking or high cholesterol. A clot forms a fissure in the plaque. The clot may shut off the blood flow of the coronary artery causing a heart attack. It is deceptively simple. The coronary arteries are clogged. No blood can flow, so the muscles of the heart cannot be supported, and heart metabolism stops leading to death.

In Germany, another theory of myocardial infarction was proposed by Dr. Berthold Kern (1911-1995). Dr. Kern, while performing autopsies in Germany in the 1930s and 1940s, observed that the findings of these autopsies did not corroborate the coronary obstruction hypothesis. He began researching the literature, looking for clues as to an alternative cause. What he found was not only a new theory that may provide the missing piece of the coronary obstruction theory but a therapy now being used by more than 5000 physicians in Germany with reportedly remarkable success.

Dr. Berthold Kern

Dr. Kern's claims, as set forth in his 1971 informational paper *Three Ways to Cardiac Infarction*, can be summarized as follows:

1. The coronary obstruction theory cannot adequately explain observed facts.

2. The major causal factor underlying heart attack is a primary chemical destructive process caused by unchecked metabolic acidosis (accumulation of acid) in the left ventricular tissue and substantially unrelated to coronary artery disease.

3. The regular, clinical use of oral g-strophathin (a cardiac glycoside derived from the West African plant strophanthus gratus): Prevents lethal myocardial tissue acidosis, and thereby substantially reduces the incidence of heart attack and completely prevents infarction deaths.

Dr. Kern's observations that most heart attacks occur in patients without significant obstruction of the coronary artery supplying the infracted tissue finds great support in the American peer-reviewed literature. Since 1948, more than a dozen reports of post-mortem examination of infarcted hearts have consistently failed to corroborate the coronary artery thrombosis theory of myocardial infarction. Many victims of fatal heart attacks have had no evidence whatsoever of coronary occlusions (blockages).

An example of the degree of non-confirmation can be ascertained by the following quote from a 1980 article in the journal *Circulation*: "These data support the concept that an occlusive coronary thrombus has no primary role in the pathogenesis of a myocardial infarct." The simple explanation of this is that although there is a blockage present, it is not the primary cause of a heat attack. The reviewer went on to note: "These reports also present clear refutation of the most common explanation used today to dismiss autopsy findings which detect no coronary thrombi, i.e., that thrombi existed at infarction but have since lysed (destruction of cells), embolized or washed away." The reviewer is mentioning that at the time of a heart attack there might be blockage of the artery, but that it is not the cause of the heart attack.

There does not appear to be any literature that effectively refutes these autopsy findings.

Another source of inconsistent data are the many reports in the literature of myocardial infarction in patients without coronary artery disease as deduced by normal coronary angiograms. Other autopsy data have revealed widely scattered areas of necrotic tissue that produces a substantial incongruence between the area of infarction and the arterial supply.

In a 1988 editorial published in the *New England Journal of Medicine* titled "Twenty years of coronary bypass surgery," Thomas Killip observed that "Neither the VA [Veterans' Administration] nor CASS [the National Institute of Health's Coronary Artery Surgery Study] has detected a significant difference in long-term survival between the two assigned treatment groups [surgical vs. medical] when all patients have been included..."[1]

More recent work with coronary angioplasty and anti-thrombolytic agents has also failed to demonstrate any clear-cut improvements in survival.

Dr. Kern went a step further. In his review of the literature, he came across the notion of collaterals, a finely-meshed network of small blood vessels that act as natural bypass channels in the heart muscle. These collaterals[2] have been made visible by Professor Giorgio Baroldi in studies at the Armed Forces Institute of Pathology. The natural collaterals in the body are similar to side streets surrounding a freeway. In this analogy, the freeway is the main arteries. When the freeway is clogged many people will take side streets to reach their destination. In the human body when the main artery is blocked, then the blood flows through the collaterals.

Baroldi developed a technique for filling the arteries of the heart with artificial blood, a chemical substance that thickens in the blood vessels. When later the tissues were dissolved in acid, the entire structure of blood vessels in the heart was revealed. Kern hypothesized that bypass grafts were created naturally by the body via the collaterals whenever a coronary artery became blocked. Therefore, heart bypass would be redundant to a large degree.

A study by Rentrop *et al* in the April 1, 1988, issue of *The American Journal of Cardiology* has produced results completely at odds with the coronary artery blockage theory but consistent with Kern's hypothesis. In an accompanying editorial, Dr. Stephen Epstein of the National Heart, Lung and Blood Institute summarizes Rentrop and colleagues' "extremely important observations." They found that in an advanced state of the narrowing of the coronary arteries, the supply of blood to the heart muscles is fully assured via collaterals that

---

[1] Killip T. Twenty years of coronary bypass surgery. *New England Journal of Medicine*, 1988 Aug 11, Vol. 319:366-368

[2] Giorgio Baroldi, *Coronary Circulation In The Normal and the Pathologic Heart*, (Washington, D.C., Office of the Surgeon General Dept. of the Army, 1967)

enlarge naturally in response to the blockage. Interestingly, it was observed that the more the coronaries narrow, the less danger there is of heart infarction. [3]

Dr. Kern's second claim, i.e., his proposed new theory of metabolic acidosis, can be summarized as follows: Metabolic conditions in the most healthy of hearts are, at best, marginal in the constantly beating left ventricle. This is the part of the heart responsible for pumping blood to most of the body, the right ventricle merely supplying the lungs. Oxygen and energy requirements are always perilously close to available supplies, and any of the several stressors may cause an oxygen/energy deficit with deterioration in oxidative metabolism and consequent development of acidosis. Lack of oxygen sets off the process of zymosis or fermentation metabolism, an anaerobic process, in order to produce energy in the cells. This, in turn, lowers the pH.

This lowering of the pH is significant as it sets off a destructive chemical process, literally a suicide reaction of the cell. Lysozymal enzymes are released, causing cell self-digestion. This starts as a single point in the muscle, then many points that eventually join to form a small area of necrotic tissue. Finally, a critical mass is reached, no bigger than the head of a pin, that triggers larger and larger areas of damaged tissue resulting in a heart attack.

Ideally then, the remedy to address infarction would be a restoration of pH balance to the heart muscle, thereby preventing tissue damage and fatal infarction. The problem Kern faced was how to accomplish this without causing positive inotropy [increasing the strength of the muscular contraction], i.e., without putting further stress on the contracting heart muscle itself. The cardiac glycosides, including digitalis and the strophanthin byproduct known as *ouabain*, is known to produce such a deleterious effect, and this is why it is not

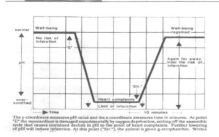

Within 10 minutes of oral Strophanthin, the pH returns to normal.

The y-coordinate measures pH value and the x-coordinate measures time in minutes. At point "E" the myocardium is damaged experimentally by oxygen deprivation, setting off the anaerobic cycle that causes continued decline in pH to the point of heart complaints. Further lowering of pH will induce infarction. At this point ("Str."), the animal is given g-strophanthin. Within

[3] Rentrop et al, *The American Journal of Cardiology*, April 1, 1988

effective against infarction.

This is where Kern made an important re-discovery. In reviewing the literature, he came across the work of Dr. Edens, who in the 1920s had reported on a qualitatively different effect of strophanthin given intravenously versus orally. Specifically, the positive inotropic effects [that is, increasing contraction] that accompanied intravenous administration were not observed with oral administration.

This important observation has been confirmed in a study by Belz published in the *European Journal of Clinical Pharmacology* in 1984. Utilizing a randomized, placebo-controlled, double blind methodology, the researchers found that the intravenous *ouabain* (strophanthin) produced the expected increase in cardiac inotropy. However, the investigators stated quite definitely that, "...the single sublingual (oral) dose of ouabain did not exert a positive inotropic effect." [4]

The most likely method of action, based on animal research done by Adams, Powell and Erdmann, is that there are two receptors in the heart: "High affinity" and "Low affinity." It is thought that intravenous administration triggers low affinity receptors and thus positive inotropy. High affinity receptors, on the other hand, react to small concentrations of g-strophanthin via oral administration thereby avoiding the dangerous effect of positive inotropy.

Dr. Kern reported results of his clinical practice in Stuttgart during the period 1947-1968 involving more than 15,000 patients. His patients treated with oral g-strophanthin experienced no fatal heart attacks and only 20 non-fatal heart infarcts. These patients included many suffering infarctions prior to entering the study. In contrast with these results, government statistics for the same time period would have predicted more than 120 fatal heart attacks and 400 non-fatal infarctions in a group of patients this size.

Currently, there are approximately 5,000 M.D.s in Germany using and prescribing oral g-strophanthin. The booklet *Eine Dokumentation ambulanz-kardiologischer Therapie Ergebnisse nach Anwendung oralen g-strophanthin* represents the results of a survey wherein 3,645 medical doctors made statements on the use of this remedy in their practices from 1976 to 1983. Of these, 3,552 gave exclusively positive testimony with no reservations. No one gave a negative response.

---

[4] Belz, *European Journal of Clinical Pharmacology*, 1984

In addition to accumulating clinical experience, a number of studies have demonstrated excellent results with oral g-strophanthin. One fascinating report in a real-life setting took place at a German coal mine. During the period 1972-1974, miners suffered episodes of acute chest pain 229 times. Medical help was a two-hour ride away, and 11 miners died during this period. From 1975-1980, all miners who experienced acute chest pain (280 episodes) were immediately given oral g-strophanthin. During this period, which was twice as long as the comparison period, no miners died after the onset of symptoms. No toxic side effects were observed. Many variables were studied, i.e., age, better access to treatment, different working conditions, etcetera to ensure comparability of observation periods.

A rigorous, double blind, randomized control study of oral g-strophanthin in the treatment of angina showed impressive results at statistically different levels. After fourteen days, 81% of patients in the treated group experienced a reduction in attacks, while in the control group, 72% receiving placebos registered an increase in attacks.

In a study of 150 seriously ill heart patients, who altogether had 254 heart attacks, oral g-strophanthin was successful in 85% of the cases. Dr. Dohrmann, who conducted the study, observed, "A positive result was registered when the severe heart attack abated at least five minutes after the g-strophanthin capsule was bitten through, and after ten minutes at the latest, they disappeared completely." [5]

A consistent feature of clinical reports using oral g-strophanthin is the absence of side effects. The cost of this remedy, which is currently available to German physicians and their patients, is approximately $30 per month for typical use. (This was in 1991.)

At this point, every indication suggests that oral g-strophanthin may be a significant breakthrough in the treatment and prevention of myocardial infarction. What is needed is a definitive American clinical trial.

At an annual meeting of the American College of Cardiology in New Orleans, it was stated that every year one million U.S. citizens

[5] R. E. Dohrmann et al: Klinisch-poliklinische Studie über die Wirksamkeit von g-Strophanthin bei Angina pectoris und Myokardinfarkt. Cardiol Bull 14/15: 183-187,1977
R. E. Dohrmann & M. Dohrmann: Neue Therapie der instabilen Angina pectoris bei koronarer Herzerkrankung Erfahrungsheilkunde 33: 183-90,1984

suffer heart attacks. Of these, about 60% get to the hospital alive. About 16% never leave the hospitals, and an additional 10% die within one year. This should be keen motivation for a complete and intensive investigation of the benefits of g-strophanthin.

The prospect of replacing heart bypass surgery with a safer, more effective and less expensive treatment will not be welcomed by those who benefit from the surgery and drugs that are a part of the heart bypass process.

There is no reason why this technique and herb should not be examined and utilized in our medical system.

In August 25, 1989, Robert Maver received the following letter (in part) from the Senior Research Associate for the University of Colorado at Denver.

"Dear Bob:

I am pleased that our preliminary report on Dr. Kern's theories fulfilled your expectations. This is an interesting project and one well worth pursuing. The relative lack of evidence in support of the orthodox theory of myocardial infarction is somewhat surprising. In fact, an article for publication could be written based on this part of the report. Would you have any objections to my writing and submitting such an article?

Bob, I want to let you know how refreshing it is to find someone in the insurance industry with the vision to seek out and investigate alternative diagnostic, preventive and treatment modalities. A sustained effort to search for and properly evaluate alternative health care measures will likely yield many valuable findings." [6]

A follow-up report on Strophanthin on January 26, 1990, from the University of Colorado contained an interesting report regarding a law suit filed in Germany. The defendant was a reporter for a newspaper "...that claimed in a popular magazine that the use of strophanthin for heart disease constituted charlatanism, in particular because this medication would increase oxygen demand and thereby have an effect 'exactly the opposite of the desired, namely an extension of the infarct.' An appeals court upheld a lower court ruling that prohibited the defendant from making any further statements of this kind, primarily

---

[6] Personal Letter from University of Colorado at Denver Center for Health Ethics and Policy to Robert Maver, V.P. Mutual Benefit Life, August 25, 1989. Given to author by Robert Maver.

on the grounds that his claims were unproven." [7] Any time and every time that the reporter made the comment he would be fined substantially for his statements.

The report went on to say, "...Another plaintiff witness for strophanthin, Dr. Draczynski, described a survey of 3,654 physicians who had used strophanthin between 1976 and 1983. Reportedly, 3,552 of these physicians were unreservedly positive about strophanthin, 93 were positive 'with some reservations,' and nobody had a negative opinion." [8]

"No defendant experts were cited in this appeals court opinion. The report referred to inquiries that had been made at the German Health Ministry; apparently the ministry had received no reports of serious 'intoxication' or death as a result of strophanthin." [9]

This herb is found only in Africa, and the little g-strophanthin pill in 1990 was retailing for about $1 per day. The cost of heart bypass surgery can reach as high as $100,000. I don't believe it is too difficult to understand why medical doctors would prefer the money from heart bypass over the money for the little pill!

Dr. Kern tried to tell us that we needed to look deeper into our hearts. He pointed us in the right direction - and nothing happened.

BUT YOU CAN MAKE IT HAPPEN!

---

[7] Center for Health Ethics and Policy, University of Colorado, Follow-up report on Strophanthin prepared for Robert Maver Vice-President, Mutual Benefit Life. Given to author by Robert Maver, January 26, 1990.
[8] IBID.
[9] IBID.

There is no religion higher than truth.

Chapter 10
Visualizing the Meridians

In 1985, I learned about a unique research project that was taking place at the Necker Hospital in Paris, France. Two nuclear medical doctors were investigating whether they could prove that acupuncture points and meridians actually existed.

I decided to travel to Paris and meet with the doctors to find out what the results were of their experiments. Before I left on my trip, I was able to find some information regarding the background of the doctors.

Both doctors had very impressive medical backgrounds. Dr. Pierre de Vernejoul, M.D., is the Chairman of the Biophysical Medical Department for Necker Hospital in Paris. He was chief of the Nuclear Medicine section of Infant Maladies at Necker. Dr. de Vernejoul was also the Chief of Service of Public Service Medicine in Paris. In addition, Dr. de Vernejoul was a member of the French Commission on Atomic Energy and held numerous positions with other groups involved with biophysical and nuclear imaging.

Dr. Jean-Claude Darras, M.D., is a nuclear medical doctor as well as President of the World Union of Scientific Acupuncture Society.

Dr. Darras was working under Dr. de Vernejoul when Darras devised a unique experiment to test the validity of specific acupuncture points and whether or not there is a mechanism of energy transport along the purported acupuncture meridians. Dr. Darras wanted to devise accurate experiments that could be recreated in other laboratories in case his experimentations yielded any tangible results.

Dr. Darras, working with his colleague Dr. de Vernejoul, decided to experiment by injecting various acupuncture points with a solution using a radioactive trace mineral while doing a high-speed cat scan.

The doctors chose Te99 as their solution. The results showed that the actual molecules of radioactive material, as well as the carrier agent, rapidly moved along the meridian directly to the organ that corresponded, according to classical Chinese acupuncture literature, to the point injected. They found that 40% of the solution was transferred within 12 seconds.

Working with 70 patients in a hospital renal ward and 50 subjects with healthy kidneys, the doctors found that the isotope consistently migrated along the same pathways. There are reportedly 12 bilateral meridian pairs and 2 medial meridians.

In their main study, Darras and de Vernejoul simultaneously injected Te99 at the point "kidney 7" on both ankles and scanned the flow up the legs. They tested the rate of flow in the kidney meridian, which was symmetrical in both legs of normal subjects. For patients with degenerative kidney problems the flow in the leg on the unhealthy side showed a scattered pattern of the Te99. (See following pictures).

The doctors also discovered that laser stimulation of the kidney points distal to the point of injection caused the flow rate to increase in both normal and degenerative conditions. The stimulation responses from the laser showed the doctors that this aspect could lead to new diagnostic techniques.

An interesting experiment showed that if a person had had a kidney removed, the flow of the kidney meridian was blocked, and the flow migrated to the nearest "healthy" meridian.

Dr. Darras

Dr. De Vernejoul

While I was visiting their research area in Necker Hospital, these fine doctors performed several experiments for me to see. What was very interesting was when they injected the body with the Te99, and it was not on an acupuncture point, the substance just pooled up and remained at that spot. When they injected at any acupuncture point, the isotope then migrated in lines corresponding 100% to the ancient Chinese meridian drawings.

The doctors were able to prove that the substance was not moving in the lymphatic, nervous or circulatory systems. When laser stimulation was used on non-acupuncture points, there were no measured isotope movements. The doctors found that the isotope moved about 3 centimeters per second along the meridian lines.

I believe that this work is very exciting. First, the doctors have been able to make visible a subtle energy that courses through the body. What other subtle energies might also be present that we are not aware of? Second, for too long, aspects of acupuncture have been ridiculed by standard medical doctors. In addition, these two doctors found that the meridians follow the same pathways as laid out by the ancient Chinese texts several thousand years ago. How did ancient Chinese writers see and understand these points and meridians?

Dr. Darras and Dr. de Vernejoul also used their research to help ascertain the effectiveness of some of the standard operations that took place in hospitals. When someone had surgery on a particular organ, the doctors were able to test the patient to see if the flow of energy was re-established in the organ that was operated upon. This would be very difficult using standard methods. The surgery might appear to have been successful from the visual standpoint of the organ being repaired, but that would not necessarily correlate with how effective the organ was working.

Although the findings of Darras and de Vernejoul have been accepted by the French Academy of Medicine, in the last twenty years I have met only a few acupuncturists who were aware of this landmark research. For that matter, I haven't met any doctors who heard of the experiments. What I still find at the time of this writing are comments from some of our leading medical information services that claim that acupuncture is a fringe medical technique that has not been proved to be effective.

Two respected doctors proved that acupuncture is effective and that meridians exist, but is this ancient method used as often as it could be—as a safe modality for healing?

Bull. Acad. Natle Méd., 1985, 169, n° 7, 1071-1075, séance du 22 octobre 1985

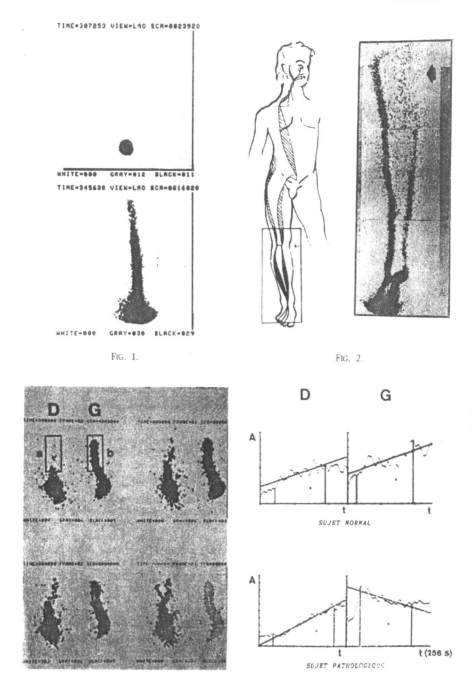

FIG. 1.

FIG. 2.

SUJET NORMAL

SUJET PATHOLOGIQUE

Measuring the flow of Te99 along meridian of legs

"Truth lies within a little circle and certain compass, but error is immense." - Henry St. John

Chapter 11
The Cast Off Cell

Why nothing has happened with the following information comes as an absolute mystery to me. This report was presented by several of the leading medical doctors in the United States. I will begin with the actual article and then give the biographical data that I found regarding one of these doctors.

"A preliminary report of a new test for **uterine and cervical cancer** was made to the Chicago Gynecological Society by S.A. F. Lash and Ralph W. Gerard of the University of Illinois, and G. Falk of the University of Chicago. The test, a simple one, involves measuring the electrical potential difference between the inside and outside of single cells cast off from the vaginal tract. The measuring technique is one that Dr. Gerard has used for many years in his basic research in nerve physiology."

"The positive and negative readings, when properly interpreted, were found to reveal whether the patient had cancer of the uterus or of the cervix. In the first series of 57 cases, the test indicated that 20 of the women had cancer; 18 of these latter were *proved* to have cancer, and two really did not. The test showed that 35 women did not have cancer, and it was right in every case. In further trials, there has never been a false negative test, and the positives were about 94 percent correct. It is hoped that the new method will make earlier diagnosis possible. However, the report stresses that the test is purely in the research stage; it will not be known for at least a year whether it will be worth using clinically." [1]

I especially find the last sentence interesting, "...whether it would be worth using clinically." This test is non-invasive and has a success rate that is statistically very high! How does a test procedure like this just disappear? This method of testing correlates well with the research work of Dr. Burr, Dr. Crile, Dr. Nordenstrom and Lakhovsky.

Here is the biographical information on one of the doctors who presented the information in 1954.

---

[1] Ralph W. Gerard, S.A. F. Lash, G. Falk, Science, Vol. 120; July 2, 1954

"Ralph Waldo Gerard devoted over fifty years of his life to scientific discovery and education. Professor Gerard received a Ph.D. in physiology in 1921 and an M.D. degree in 1924 from the University of Chicago. He was in the Department of Physiology at the University of Chicago from 1923 until 1952. He left the University of Chicago in 1952 to become director of laboratories at the Neuropsychiatric Institute of the University of Illinois. In 1955 Professor Gerard moved to the University of Michigan where he helped to found the Mental Health Research Institute. He served as director of laboratories and professor of neurophysiology in the Department of Psychiatry and Physiology."

"Professor Gerard was an uncommon scientist. He published over 500 scientific papers and nine books. He was clearly one of the most productive and distinguished neurobiologists of this century. Through his un-resting pursuit of fundamental knowledge he taught us how cells work and he helped us to understand the organization and integrative functions of nerve cells. His studies in neurophysiology included investigations of the electrical activity of the brain during sleep and the nature of brain waves. In order to examine the electrical activity of single nerve cells he developed microelectrode-recording procedures. This technical development revolutionized research in neurobiology." [2]

Dr. Gerard

Dr. Gerard had so many awards and honors that the listing of these would take several pages.

I have included this information because it is a mystery to me that this particular aspect of Dr. Gerard's research ends with his presentation to the society, and there is no further reference made to this line of investigation.

Obviously Dr. Gerard's work has had an influence upon numerous medical subjects. He presented his research regarding the electrical aspects relating to the cancer situation - and nothing happened.[3]

---

[2] Biography from University of California, 1976
[3] See note 1.

A preliminary report of a new test for uterine and cervical cancer was made to the Chicago Gynecological Society by S. A. F. Lash and Ralph W. Gerard of the University of Illinois, and G. Falk of the University of Chicago. The test, a simple one, involves measuring the electric potential difference between the inside and outside of single cells cast off from the vaginal tract. The measuring technique is one that Dr. Gerard has used for many years in his basic research in nerve physiology.

The positive and negative readings, when properly interpreted, were found to reveal whether the patient had cancer of the uterus or of the cervix. In the first series of 57 cases, the test indicated that 20 of the women had cancer; 18 of these latter were proved to have cancer, and two really did not. The test showed that 35 women did not have cancer, and it was right in every case. In further trials, there has never been a false negative test, and the positives were about 94 percent correct. It is hoped that the new method will make earlier diagnosis possible. However, the report stresses that the test is purely in the research stage; it will not be known for at least a year whether it will be worth using clinically.

An accelerated program to study all aspects of the radiation sterilization of foods is being undertaken by the Army Quartermaster Corps for the benefit of the Armed Forces. The 5-yr program will have the active participation of the Atomic Energy Commission, elements of the Armed Forces, and other governmental

"Come forth into the light of things; let nature be your teacher." - Wordsworth

Chapter 12
Sam the Eagle

In the late 1980s, while I was driving to my office in Los Angeles, I heard a very interesting news report on one of our local radio stations. The story was about an eagle that had a severely injured wing that before then no one had been able to repair. The eagle, that had been given the name Sam, had been to some top veterinarians who were not able to help. Finally, Sam ended up at the office of a certain vet who lived in a city just outside of Los Angeles. The vet used a special device that accelerated healing, and the broken wing was now mended. The machine gave off electronic waves.

Of course, this intrigued me and, I attempted to locate the doctor who lived in the Simi Valley. I was finally able to reach the offices of the doctor, but I was told that he was too busy to speak with me. The question I was asked was whether I was calling about a pet that I owned or was I just calling to get information about the healing machine.

When I mentioned that I was interested in learning about the technology, I was politely told that the doctor was not speaking to anyone, and the machine could never be used to help people. That was an interesting response because I had not stated that I was looking to treat someone.

I continued to call back and mentioned that I was working for a non-profit organization, and I was familiar with what the doctor might be experiencing. I finally received a phone call from the doctor late one evening. I shared with him that I had offices in Europe and was sure that he was using a device that was from out of the U.S., and I also was sure that he had received telephone calls from medical agencies here in the U.S. After I mentioned that, he invited me to come to his offices the following day.

The doctor was utilizing a special machine from Europe. He had been using the machine on many of his animal patients with fantastic success. The machine used pulsed electromagnetic frequencies to accelerate bone and tissue healing. The doctor shared with me how excited he and his staff were because they had healed an animal that the local zoo personnel and other vets just could not help. However, the doctor did not expect what the response would be from the public.

Instead of people calling about just their pets, they were calling and asking if the machine could help their friends and relatives who had various medical and health conditions. At first, the doctor was answering in the affirmative because cells are the same whether they are in an animal or a human being. And the action is the same regarding the healing effects of the machine. The doctor then related that he had been paid a visit by government authorities who warned him if he made any comments regarding this machine or technology being able to help people that he would never practice any type of medicine again.

He told me it was enough to scare him worse than he had ever been frightened in his life. If you are shocked or do not believe what I am saying then consider this. This doctors experience is not an unusual case. In my files I have other letters from researchers who have been threatened during the years and told not to discuss electromagnetic devices being utilized on humans. You will be surprised when you read the medical reports, studies and literature that are coming out of the veterinary community. Various health problems that are affecting humans are also found in the animal population including arthritis, lameness and Sam's difficulty with healing, and other health challenges are addressed by machines and devices that are not available for humans. These machines use some of the technologies that I have shared with you in this book. Now why should they be available for animals but not for people? The structure of animal cells is the same as human cells, so health problems are fundamentally the same in humans and animals. Furthermore, adverse effects are not showing up in the animals after electromagnetic treatments, and no deaths have been reported from these machines.

As I am writing this section, a report has become public that a trial of a new drug for diabetes has been stopped because 273 people have died as the trial is progressing. The FDA has stopped the testing. The article is very strange as it reads that other trials for drugs for diabetics normally average only 203 people dying during the trial. I think this is disgusting. So many people have been upset over animal testing, and yet we all seem to accept the people testing that is taking place.

Electromagnetic technology is here - and nothing happens for us humans.

BUT YOU CAN MAKE IT HAPPEN!

"Wonder at all things before you, for wonder is the beginning of knowledge." -
The Oxyrynchus Papyri

Chapter 13
The Cure That Time Forgot: Ultraviolet Blood Irradiation

Dr. Rowen

Robert Jay Rowen, M.D., wrote the information for this chapter. Permission to reprint this excerpt has been given by Dr. Rowen. I have replaced some of the more highly technical medical terms for easier reading.

In the 1940s, a multitude of articles appeared in American literature detailing a novel treatment for infection. This treatment had a cure rate of 98 to 100% in early and moderately advanced infections and approximately 50% in terminally moribund (dying or near dead) patients. Healing was not limited to just bacterial infections but also viral (acute polio) wounds, asthma and arthritis. Recent German literature has demonstrated profound improvements in a number of biochemical and blood tissue markers. There has never been any toxicity, side effects or injury reported except for occasional inflammation.

Ultraviolet (UV) light has been known for decades to have a sterilizing effect and has been used in many different industries for that purpose. Almost all bacteria may be killed or severely weakened by ultraviolet rays, but there is considerable variation in the rapidity of bacteria destruction. Those that live in the body are most easily affected, while those in nature adapt to the action of sunlight and become relatively resistant to radiation. [1] LJV-sensitive bacteria have not been shown to become resistant, and toxins have been found to be very unstable in the presence of UV irradiation (Diphtheria, tetanus, and snake venom are inactivated by ultraviolet rays.) [2]

At the turn of the century, Niels Finson was awarded the Nobel Prize for his work regarding UV rays and various skin conditions that showed a success rate of 98% in thousands of cases, mostly lupus

---

[1] Henry Laurens, The Physiological Effects of Ultraviolet Irradiation, JAMA, Vol. 11, No. 26, December 24, 1938

[2] Henry Barrett, "The Irradiation of Autotransfused Blood by Ultraviolet Spectral Energy: Results of Therapy in 110 Cases", *Medical Clinics of North America*, May, 1940, pp. 723-732.

vulgaris. [3] Walter Ude reported a series of 100 cases of Erysipelas in the 1920s, claiming a nearly 100% cure rate with UV skin irradiation. Emmett Knott pioneered the irradiation of autologous blood on dogs before treating a moribund woman (near death) with post-abortion sepsis in 1933. She was thought to be untreatable. With his treatment of blood irradiation, the woman promptly recovered resulting in more research and further development of the "Knott" Technique. [4] The technique involved removing approximately 1.5cc/per body pound, citrating it for antoagulation and passing it through a radiation chamber. Exposure time per given unit of each cc was about ten seconds.

By the early 1940s, UV blood was being used in several American hospitals. Into the late 1940s, numerous reports were made about the use of UV blood irradiation and its complete safety and efficacy is keeping infections to a minimum. With the emergence of antibiotic therapy, the reports suddenly ceased. In the ensuing years, German literature demonstrated the effectiveness of UV irradiation in vascular conditions. Additionally, more thorough observations of significant improvement in many physiologic processes and parameters have been reported.

The most prolific American researcher of UV blood irradiation was George Miley, a clinical professor at Hahnemann Hospital and College of Medicine, who practiced the Knott technique at the hospital's blood irradiation clinic. In 1942, he reported on 103 consecutive cases of acute pyogenic infections at Hahnemann Hospital in Philadelphia. Such conditions included puerperal sepsis, sinusitis, pyelitis, wound infections, peritonitis (ten cases) and numerous other forms of infections. Results of recovery were 100% for early infections, 46 of 47 for moderately advanced, and 17 out of 36 of those who were near death. [5] Staphylococcus had a high death rate, but those patients were using sulfa drugs, which may have inhibited the effectiveness of the UV radiation treatments. In fact, when Miley

---

[3] G. Frick, "A Linke: Die Ultraviolet bestrahlung des Blutes, ihre Entwicklung und derzeitiger Stand"., *Zschr-artl.*, Forth. 80, 1986.

[4] Emmett Knott, *Development of Ultraviolet Blood Irradiation*, American Journal of Surgery, August 1948, pp. 165-171

[5] George Miley, *The Knott Technique of Ultraviolet Blood Irradiation in Acute Pyogenic Infections*, The New York State Journal of Medicine, January 1 1942, pp. 38-46

reviewed his data, he found that all the Staph failures had been on sulfa first. A second series of nine patients (six Staph aureus, three Staph albus) had a 100% recovery rate with one or two treatments when sulfa was not used. [6]

Rebbeck and Miley documented the fever curve of septicemia in patients who received UV therapy, demonstrating detoxification and recovery within a few days. [7] In 1947, Miley reaffirmed his initial finding reporting on 445 cases of acute pyogenic infection, including 151 consecutive cases. Again, results showed a 100% recovery in early cases (56), 98% recovery in moderately advanced cases (323) and 45% in apparently near death patients (66). [8] Detoxification usually began within 24 to 48 hours and was complete in 46 to 72 hours. Some patients required only one or two irradiation treatments, while a few needed one or two more.

Effectiveness in other viral conditions was further documented by Olney. [9] His report documented 43 patients with acute viral hepatitis treated with the Knott Technique. Thirty-one patients had acute infectious hepatitis; 12 had acute serum hepatitis (hepatitis B). An average of 3.28 treatments per patient were administered; the average period of illness after the treatment was 19.2 days. Two recurrences were observed among the 43 patients during a follow-up period averaging 3.56 years, one in each type of hepatitis. The one suspected recurrence in the "serum" variety was in a heroin addict, and re-infection was suspected. No deaths occurred among the 43 patients during the follow-up period. Marked improvement and rapid subsidence of symptoms was noted in all patients treated within three days or less in 27 patients. 11 showed marked improvement in 4 to 7 days, and five patients showed improvement in 8 to 14 days.

Rebbeck reported a remarkable effect on the autonomic nervous system, documenting how post-surgical paralytic intestinal obstruction could be relieved very quickly with LTV blood irradiation. [10] He

---

[6] Miley and Christensen, *Ultraviolet Blood Irradiation Therapy: Further Studies in Acute Infections*, American Journal of Surgery, Vol. 73, No. 4, April 1947, pp. 486-493

[7] Rebbeck and Miley, Review of Gastroenterology, January-February, 43. p.11.

[8] Miley and Christensen, *Ultraviolet Blood Irradiation Therapy: Further Studies in Acute Infections*, American Journal of Surgery, vol 73, No.4, April, 1947, pp. 486-493.

[9] Olney, R.C., American Journal of Surgery, Vol. 90, September 1955, pp. 402-409

[10] Rebbeck, E.W., Review of Gastroenterology, January-February, 1943.

attributed this effect to toning the autonomic nervous system. Autonomic effects can also be appreciated in the reports on asthma.

The authors were so impressed with the results that they included numerous case reports of hopeless and long-suffering infectious conditions being resolved with UV blood irradiation. Rebbeck reported on its prophylactic preoperative use in infectious conditions, concluding that the technique provided significant protection with a marked decrease in morbidity and mortality. [11] The remarkable lack of any toxicity was consistently noted by all authors. In addition to polio, Miley reported that viruses in general responded in similar fashion to the infections with pus present. [12] Botulism, often fatal even today, was treated by Miley. [13] A botulism patient was in a coma and could not swallow or see. Within 48 to 72 hours of one irradiation treatment, the patient was able to swallow, see and was mentally clear. She was discharged in excellent condition in a total of 13 days. UV blood irradiation resulted in prompt healing of chronic very long-term, non-healing wounds.

Miley went on to discuss a summary of physiologic changes documented through the 1940s including inactivation of toxins and viruses, destruction and inhibition of growth of bacteria, increase in oxygen-combing power of the blood, activation of steroids, increased cell permeability, absorption of ultraviolet rays by blood and emanation of secondary irradiations (absorbed UV photons re-emitted over time by re-perfused blood), activation of sterols into Vitamin D, increase in red blood cells and normalization of white cell count.

Recent German research reports significant improvement in vascular conditions when using ultraviolet blood irradiation, including peripheral arterial disease and Raynaud's disese. One study demonstrated a 124% increase in painless walking with patients with

[11] Rebbeck, E.W.,Preoperative Hemo-Irradiatiotts, American Journal of Surgery, August, 1943, pp. 259-265.

[12] Miley and Christensen, Ultraviolet Blood Irradiation Therapy in Acute Virus and Virus-Like Infections, The Review of Gastroenterology, Vol. 25, No. 4, April 1948, pp. 271-276.

[13] Meiley George, Recovery From Botulism Coma Following Ultraviolet Blood Irradiation, The Review of Gastroenterology, Vol. 13, No. 1, January-February, 1946. pp. 17-18.

Stage IIb occlusive disease (Fontaine) as compared to 48% improvement with pentoxifylline. [14]

In the 1800s, arguments raged between Pasteur and his rival Bechamp over the true cause of infectious disease. Pasteur claimed the cause was the organism alone, while Bechamp claimed the disease rose from organisms already within the body that had pleomorphic capability (the ability to change as noted by Royal Rife who watched it take place through the Rife Universal Microscope). It is rumored that Pasteur on his deathbed admitted that Bechamp was correct.

Modern medicine has focused on drugs to suppress symptoms or inhibit certain physiology (NSAID drugs) to treat disease. As a result, we have seen the frightening rise of resistant organisms and the side effects of chemical pharmacology. Perhaps medicine should consider the concept of non-specific modalities that encourage the body's healing response and immune system. What could be safer or more effective against infection than the bacteriocidal capabilities of our own phagocytes and a properly functioning immune system?

"Ultraviolet irradiation of blood has been approved by the FDA for the treatment of cutaneous T-cell lymphoma. Thus, the method is legal in the context of the FDA's definition of legality. It is also legal, from the standpoint of long (over 50 years) and continuous use by physicians in the United States as a commercially viable product before the present FDA was even in existence." [15]

This simple, inexpensive and non-specific technique was clearly shown years ago to be a totally safe and extremely effective method of treating and curing infection, promoting oxygenation, vasodilation, improving asthma, enhancing body physiology and circulation and treating a variety of specific diseases. Its use in hospitals and offices could significantly reduce mortality, morbidity and human suffering. Much more research needs to be done in determining all of the potential uses of ultraviolet blood irradiation therapy and also its correlation with other oxidative therapies.

To contact Dr. Rowen please use the following internet addresses:

Robert Jay Rowen, MD
www.secondopinionnewsletter.com or www.doctorrowen.com

---

[14] Pohlmann, et al, Wirksamkeit Von Pentoxifyllin und der Hamatogenen Oxydationstherapie, Natur-und GanzheitsMedizin, 1992; 5:80-4.
[15] Weg, Stuart, MD Private Communication, January, 1996.

Chapter 14
Have We Left A Stone Unturned?

I have mentioned Robert W. Maver, F.S.A., M.A.A.A., of Mutual Benefit Life Insurance a number of times in previous chapters. He was an open-minded visionary. This chapter is basically his chapter wherein he questions whether the insurance industry should participate in finding safer, more effective and cost-efficient medical therapies. This chapter reproduces a paper he wrote to provoke a thoughtful and intelligent approach to research and new therapies. (All paragraphs within double quotation marks are from his paper.)

"Suppose for a moment that the year is 1928, and a patient is suffering from a virulent, life-threatening bacterial infection. An independent researcher has discovered a new drug, based on a common mold, that he claims is effective in combating such infections. But the scientific establishment is ridiculing and rejecting the researcher's claims. Should the patient be allowed to take a chance on this new therapy?"

"We know that Alexander Fleming's discovery of penicillin was ridiculed and ignored for more than 12 years.[1] Later, it was proved effective and along with its derivatives has saved literally millions of lives - despite the early derision. Fleming was knighted and received a Nobel Prize."

Dr. A. Fleming

"But what if the conventional wisdom had prevailed and Fleming's drug had never gotten out of the laboratory? Impossible? An isolated and unfortunate incident in the history of science? Perhaps not."

"While most of us not involved in scientific research assume that progress occurs naturally and discoveries or breakthroughs are at once communicated and embraced by the scientific community, those who study scientific progress conclude otherwise."

"Thomas S. Kuhn in his book *The Structure of Scientific Revolutions* reviews many examples of major advances in science introducing the concept of paradigm shift. He rejects the idea of progress in science coming through accreditation. Instead, he asserts

---

[1] Moss, Ralph W., *The Cancer Syndrome*, New York Grove Press, Inc., 1980

major progress is characterized by a revolutionary process in which an older theory is rejected and replaced by an incompatible new one.[2] Typically, breakthroughs involve bitter conflict between the independent scientific innovator and the established orthodoxy of the day that seeks to discredit and combat the perceived heresy. Deviance from approved practices and doctrine is viewed as an ideological threat and not well tolerated."[3]

"The citations of science resisting new ideas are numerous. We are probably all familiar to some extent with the ideological resistance to Galileo and Copernicus and later to Darwin. Mendel's theory of genetics was largely ignored for 35 years and his work dismissed as that of a dilettante by the leading scientists of his day. [4] Ohm's law and Thomas Young's wave theory of light were similarly resisted. Harvey's monumental work on the general theory of circulation of the blood was forbidden to be taught at the University of Paris Medical School, twenty-one years after its publication."[5]

"These cannot be dismissed as historical episodes not relevant to twentieth century science, as evidenced by the reception given Alfred Wegener's 1922 theory on continental drift. As early as 1928, geologist Arthur Holmes had advanced the supporting theory of convection currents as the cause of the continental drift. Accepted by geologists today, it was resisted for forty years by the leading scientists in the field (e.g., Harold Jeffreys in England and Maurice Ewing in the U.S.)."[6]

"As biologist Hans Zinsser has commented: 'That academies and learned societies are slow to react to new ideas is the nature of things ...the dignitaries who hold high honors for past accomplishments do

[2] Kuhn, Thomas, *The Structure of Scientific Revolutions*, Chicago, Chicago University Press, 1970.
[3] Houston, Robert G., *Repression and Reform in the Evaluation of Alternative Cancer Therapies*, New York, 1987
[4] Barber, Bernard, *Resistance by Scientists to Scientific Discovery*, Science, Vol. 134, pp. 596-602, 1961
[5] Downs, Robert B., Landmarks in Science, Littleton, CO, Libraries Unlimited, Inc., 1982
[6] Broad, William and Wade, Nicholas, *Betrayers of the Truth*, New York, Simon and Schuster.

132

not usually like to see the current progress rush too rapidly out of their reach.'"[7]

"Max Planck, father of quantum physics, which has completely revolutionized science, has expressed his views in the most straightforward fashion, suggesting that new scientific truth does not triumph by convincing its opponents and making them see the light, but rather because its opponents die, and a new generation grows up that is familiar with it."[8]

"Another fascinating example that received a great deal of attention in the 1960's was the so-called Velikovsky affair. In 1950, Immanuel Velikovsky published the book *Worlds in Collision*. In it he proposed that catastrophic events as recorded in the Old Testament, the Hindu Vedas and Roman and Greek mythology were due to the earth repeatedly passing through the tail of a comet during the period of the fifteenth through seventh century B.C. Further, it was theorized that there was a near collision between Earth and Mars.[9] Reaction to the theory by leading astronomers was to dismiss the work out of hand as nonsense. Many did this without ever reading the book. Pressure was so intense that Macmillian Publishing that had originally published the book gave up the rights. The company had been threatened with a boycott by the established medical scientific community. Seventy percent of Macmillian's business at the time was in textbooks."[10]

"The treatment of Velikovsky engendered a series of articles in 1963 in *The American Behavioral Scientist* whose editor Professor Alfred de Grazia commented: 'If the judgment of the authors is correct, the scientific establishment is gravely inadequate to its professional aims, commits injustices as a matter of course, and is badly in need of research and reform.'"[11]

"De Grazia went on to publish a book in 1966 recounting the affair. In the forward, he states: 'What must be called the scientific

[7] Zinsser, H., *As I Remember Him: The Biography of R.S.*, Boston, Little Brown and Co., 1940.

[8] Planck, M., *Scientific Autobiography*, F. Gaynor, translator, New York, Philosophical Library, 1949.

[9] De Grazia, Alfred, The Velikovsky Affair—Warfare of Science and Scientism, New Hyde Park, New York, University Books.

[10] Ibid.

[11] Polanyi, Michael, *Knowing and Being*, Chicago, University of Chicago Press, 1969.

establishment rose in arms, not only against the new Velikovsky theories but against the man himself. Efforts were made to block dissemination of Dr. Velikovsky's ideas, and even to punish supporters of his investigations. Universities, scientific societies, publishing houses, the popular press were approached and threatened; social pressures and professionals sanctions were invoked to control public opinion.'" "The issues are clear: 'Who determines scientific truth? Who are its high priests, and what is their warrant? How do they establish their canons? What affect do they have on the freedom of inquiry and on public interest? In the end, some judgment must be passed upon the behavior of the scientific world and, if adverse, some remedies must be proposed.'"[12]

"Michael Polanyi discusses the Velikovsky affair in his essay on The *Growth of Science in Society*. He recounts how Velikovsky was bitterly attacked by distinguished astronomers who admitted they had not read his book. He tells of new information coming from Mariner II probes confirming some of Velikovsky's earlier predictions but still a refusal to look at his work by scientists of the day. He appreciates de Grazia's concern that perhaps new ideas in science do not gain acceptance in a rational way based on factual evidence but based instead on either random chance, ruling power, economic or political interests or as dictated by dogma."[13]

"However, Polanyi then goes on to defend science pointing out that it is not open-minded and with good reason. The conservatism of institutionalized science and its protection of currently accepted theories and 'facts' is defended on the grounds that to do otherwise would jeopardize the credibility and very survival of science against false claims. In discussing unconventional scientific claims, he goes as far as to suggest: 'Science cannot survive unless it can keep out such contributions and safeguard the basic soundness of its publications. This may lead to the neglect or even suppression of valuable contributions, but I think this risk is unavoidable.'"[14]

"There is a basic conflict here between preserving the structure, order and credibility of science and maintaining openness to new theories and anomalous data. This conflict is explored by Thomas Kuhn in another of his books, aptly titled *The Essential Tension*, where

---

[12] de Grazia.
[13] Polanyi
[14] Ibid.

he observes: 'The successful scientist must simultaneously display the characteristics of the traditionalist and of the iconoclast.'"[15]

"Marcello Truzzi in his discussion *On the Reception of Unconventional Scientific Claims*, a symposium convened by the American Academy for the Advancement of Science, suggests that it may be even more difficult today for new ideas to break through due to the escalating economics of research. Truzzi observes that: 'Unconventional ideas in science are seldom positively greeted by those benefiting from conformity.'[16] He predicts new forms of vested interests resulting from today's institutionalization of science through research programs that compete for massive funding and acknowledges '...this has been becoming a growing recognized problem in some areas of modern science.'"[17]

"Medical science has no special immunity in the area of resistance to new ideas, and its history is replete with examples like Fleming."

"In 1848 Ignaz Philipp Semmelweis, a graduate of the University of Vienna's medical department, introduced a revolutionary new procedure as assistant in the Vienna obstetrical clinic - he required students to wash their hands in chlorine water before entering the clinic. There was an immediate and dramatic decrease in the high mortality rate from puerperal fever."[18]

"Semmelweis became a staunch and vocal advocate of this procedure pleading with obstetricians to tend to patients only after aseptically clean. He was viciously attacked by his profession and dismissed from his hospital in Vienna. For the next ten years he amassed evidence to demonstrate that his antiseptic procedure would prevent death from 'childbed fever.' In 1861, he published his findings in a book, which he distributed to the major medical societies throughout Europe. He was completely ignored despite the fact that puerperal fever was ravaging maternity hospitals throughout Europe."

"The Vienna Hospital where he had eliminated death from puerperal fever in 1848 and from which he was dismissed saw 35 out

---

[15] Kuhn, Thomas S., The Essential Tension, Chicago, University of Chicago Press, 1977.
[16] Truzzi, M., *The Reception of Unconventional Science*, AAAS Selected Symposia, series-25, Seymour H. Mauskopf – editor, Boulder, CO, Westview Press, 1979.
[17] Ibid.
[18] Rich, J., *The Doctor Who Saved Babies, Ignaz Phillip Semmelweis*, New York, Messner.

of 101 patients die in autumn of 1860. In Stockholm, in the same year, 40% of all women patients got the fever with 16% dying."[19]

"Semmelweis could not cope with the knowledge that many women were dying unnecessarily. He died in an insane asylum in 1865 just prior to Lister's emergence on the scene, which will be discussed later. Today Semmelweis is credited with a major breakthrough as the first to eliminate puerperal fever."

"A similar incident occurred in the United States to no less a figure than Oliver Wendell Holmes."

"When male doctors began to replace midwives in the early nineteenth century, using forceps and making frequent examinations during labor, all without sanitation, there was an immediate and rapid increase in *child-bed fever* almost always resulting in death. This advent of disease connected with childbirth in hospitals almost destroyed the new profession of obstetrics."

In 1842 in the United States, puerperal fever was prevalent in Boston, and Oliver Wendell Holmes, then professor of anatomy at Harvard University, wrote and delivered a paper on "The Contagiousness of Puerperal Fever," the essence of which was a recitation of the evidence (dating back to 1773) supporting puerperal fever as a contagious disease transmitted by a doctor or nurse and due to specific infection."

Oliver W. Holmes

"Reaction to Holmes was swift. Those who had taught for years that puerperal fever was not contagious (the prevailing medical wisdom) began to attack him. Because Philadelphia was at the time the center in America for the teaching of obstetrics, two of the biggest names from Jefferson Medical College and the University of Pennsylvania, were the focus of the counterattack. The counterattack included such insights as, "I prefer to attribute them (puerperal fever attacks) to accident, or Providence, of which I can form a conception, rather than to a contagion of which I cannot form any clear idea, at least as to this particular malady.'"[20]

"Some 12 years later, Holmes' second edition of his original work appeared. Eventually his arguments became accepted facts in the

[19] Broad and Wade.
[20] Downs.

medical profession and saved countless number of women from untimely deaths."

"Medical historian Dr. Henry R. Viets called Holmes' paper '...the most important contribution made in America to the advancement of medicine.'"[21]

"Holmes himself said, 'I had to bear the sneers of those whose position I had assailed, and, as I believe have at last demolished. Others had cried out with all their might against the terrible evil before I did, and I gave them full credit for it. But I think I shrieked my warning louder and longer.'"[22]

"Joseph Lister is acknowledged today as ushering in a new era in medicine. In 1867 his paper *On the Antiseptic Principle in the Practice of Surgery* was read before the British Medical Association in a meeting in Dublin. He wrote, 'Since the antiseptic treatment has been brought into full operation, my wards though in other respects under precisely the same circumstance as before, have completely changed their character; so that during the last nine months not a single instance of pyemia, hospital gangrene or erysipelas has occurred in them.'"[23]

"Lister's great discovery was greeted by violent criticism and opposition by leading surgeons of the day. In fact, at an 1869 meeting of the British Medical Association, the address on surgery was devoted to an attack on the antiseptic theory. The London surgeons did not appreciate a provincial from Scotland suggesting to them how to improve surgical protocol."[24]

Dr. Lister

"Louis Pasteur's germ theory met with violent opposition from the medical community of the day by whom he was considered an outsider, a mere chemist poaching on their scientific preserves, not worthy of their attention."[25]

"Part of the reason for resistance to new ideas seems to be that many of the major advances in an area of science have come either from outsiders (scientists from another discipline) or from those not members of the recognized scientific elite. Perhaps it is no accident

[21] Ibid.
[22] Ibid.
[23] Ibid.
[24] Wrench, G.T., *Lord Lister—His life and work.*, London, T. Fisher Unwin., 1913.
[25] Barber, Bernard

that advances come from those not blinded by adherence to the established dogma of the times."

"The history of medicine reveals that independent researchers have been pivotal in determining medical progress, and yet they were usually denounced by the authorities of their day."

"Today, a new generation of researchers is providing statistical and clinical information on innovative therapies for the scourge of modern society, cancer. But in this day and age, in addition to the ridicule of the medical orthodoxy, a rigid system of testing and approval, calcified by the same suspicions of alternative therapies that plagued Fleming, can keep some of these treatments from even the most desperately ill patients who might benefit from them."

"There are many obvious reasons why we are so slow to make and utilize new treatments and remedies available to patients, chief among them our concern for patients' safety. More and more, Americans are holding the government, the business world and the medical profession to a higher standard of safety, and they are willing to use the legal system to punish any institution or individual they perceive to be violating that standard."

"However, other practical difficulties stand between patients and many of these new therapies. For example, many alternative therapies are natural remedies or processes that can't be patented. Without such patent protection, drug companies do not have the incentive to spend the millions of dollars required to develop and market the substances involved - including the estimated \$100 million[26] required to get a new drug through the FDA approval process."

"This is not an insignificant hurdle. In fact, quite a bit of research has been carried out in the last two decades confirming anti-tumor activity in plants and herbs that often are the basis of alternative therapies. But with little prospect of patentability, no industry has rushed in to finance the FDA approval process."

"Some suggest that these practical concerns are interwoven with a bias against natural biological remedies.[27] Conventional cancer treatment today consists of radiation, surgery and chemotherapy - all modalities that could be described as aggressive, often with debilitating side effects. In many cases, the treatment is feared as much as the disease itself. *Most alternative therapies on the other*

---

[26] Laszlo, J., *Understanding Cancer*, New York, Harper and Row.
[27] Moss, 1980 and Houston, 1987)

138

hand are non-invasive, non-toxic approaches. Thus, the ideological threat is heightened by a legitimate competitive economic concern from industries affiliated with conventional treatments. They are concerned that patients would try these in favor of less severe treatments."

"Ultimately, these barriers undermine not only the ability of a new treatment's advocates to gain acceptance from the scientific and medical community but also the ability to secure approval by the Food and Drug Administration."

"The entire subject of experimental therapies raises a set of complex and challenging issues for the insurance companies. We cannot ignore the fact that the illness in question is extremely costly, not only for our industry, but for society. The direct and indirect costs of fighting cancer have been estimated at one hundred billion dollars a year.[28] This cannot be far off when one considers the loss of productivity and earning power for the one million Americans stricken with cancer each year, and the estimated 500,000 who will die each year."[29]

"We must acknowledge the need for new approaches. In a 1986 article published in the *New England Journal of Medicine*, John C. Bailar, Ph.D, M.D., of the Harvard School of Public Health, and Elaine Smith, Ph.D., M.P.H., from the University of Iowa Medical Center assessed the overall progress of cancer since 1950. Bailar, an eminent biostatistian, was editor-in-chief of the National Cancer Institute (NCI) Journal for 1974-1980. The data used in the study was gathered from the National Center for Health Statistics and the SEER (Surveillance, Epidemiology and End Results) program, which was developed under the auspices of the NCI. The article concluded, 'We believe that mortality rates, age-adjusted to current standards are the best single measure of overall progress. Age-adjusted mortality rates have shown a slow and steady increase over several decades, and there is no evidence of a recent downtrend. In this clinical sense we are losing the war against cancer. The main conclusion we draw is that some 35 years of intense effort focused largely on improving treatment must be judged a qualified failure.'"[30]

---

[28] Moss
[29] Laszlo
[30] Bailar, John C. and Smith, Elaine M., *Progress Against Cancer?*, The New England Journal of Medicine, Vol. 314, No. 19.

"We have the ability to banish new treatments, and we should also recognize our influence in shaping medical practice in this country. We have the ability to banish new treatment possibilities by relegating them to the experimental clause in our contracts and denying claim payment, exacerbating some of the economic problems cited above."

"But let's get to what should be the most basic issue of our industry - can alternative therapies reduce our claim costs? Perhaps a specific case may help to put into perspective the possibilities. The case is unusual in that the patient maintained contact with his team of doctors (a prominent Head and Neck Oncology Group in New Jersey) after conventional treatment had failed. The team monitored his progress on an alternative therapy and documented complete remission attributable to that therapy as opposed to a residual effect of their own prior treatments. Moreover, they are now introducing the therapy to other patients in their practice."

"W.M., a 61 year old male, was diagnosed in March, 1985, with cancer of the mouth. During the next three and one-half years W.M. received extensive radiation treatment, had surgery four times including removal of all teeth, a removal of lymph nodes, removal of his jaw bone, part of his tongue, a section of his throat and had seven intensive in-hospital chemotherapy sessions that led to side effects that were almost fatal."

"The result of this conventional treatment was not successful. W.M. had a tumor that was now the size of a fist. His oncologists had no options except extraordinary radical surgery which he refused on quality of life grounds. Prognosis: terminal. Total cost was $81,680."

"W.M. pursued an alternative treatment involving a combined nutritional and herbal approach. The result was complete remission in six months. W.M. continues to be cancer free and enjoys a remarkable quality of life. Total cost of the alternative treatment was $2,500 (including travel costs)."

"Ironically, W.M.'s insurance policy covered the $81,680 but not the $2,500."

"The bottom line is that insurance companies cannot ignore the possibility that by helping to clear some of the obstacles that are blocking or slowing alternative therapies, we might be able to save substantial claim dollars - and that is in *our* economic best interest."

"With so much at stake, it may be well worth considering financial involvement by insurance companies in identifying promising alternative therapies and in making them available to patients, if the risks involved are carefully balanced with proper structure and appropriate standards and review."

"A program underwriting research or financing the distribution of alternative therapies to patients with otherwise untreatable illnesses could make a crucial difference in overcoming barriers to such therapies."

"Is it possible that in our desperate search for cost containment we have overlooked this rather basic concept of competition from alternatives in medicine, assuming that 'science' would handle that aspect for us? Given the potential payback, and the thousands of W.M.'s out there all telling the same story, investigation by insurance companies in this area could be time and money well spent." (W.M. was Robert Maver's father--Author).

Robert Maver passed away in December of 1997. I first met Bob in December of 1988; we had a very close and productive friendship. Although Bob was a quiet and unassuming individual, in January of 1989, he was instrumental in instituting a special research division for Mutual Benefit Life Insurance Company (MBL). At the time, MBL was the 13th largest insurance company in the Unites States.

This special research division investigated new and promising alternative treatments and therapies from around the world. Bob also had MBL reimburse many of its insureds for alternative therapies that they had taken but MBL had previously refused to pay for.

Bob's influence with the highest corporate officers within MBL allowed him to come to the support of alternative medical practitioners who were being sued by other insurance companies. Letters were sent from the executive offices of MBL directly to the corporate officers of other insurance companies challenging their actions. Imagine one insurance company actually putting pressure on other insurance companies in support of alternative medicine!

Bob traveled with LaVerne Boeckmann and me to clinics and practitioners in Europe and throughout the U.S. investigating anyone who was demonstrating successes within the alternative health field.

Bob had a great sense of humor, and I had some wonderful experiences traveling with him on business trips and spending time with him on a social level with his wife Sherri and their three children. Bob died in his mid-40s but accomplished so much.

Chapter 15
Conclusion

Do you remember in the not too distant past how virtually every other advertisement on our television was for one of the tobacco companies? Those advertisements showed happy and vibrant people living life with gusto and smiling as they were puffing away.

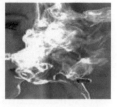

Beautiful actors showed us that having a cigarette in your hand just made life so fulfilling! And what were some of our government and non-profit health groups doing? They were doing and saying nothing. As early reports came out about possible links to cancer, we saw spokespersons for the various agencies talking while holding cigarettes in their hands. The FDA did nothing because cigarettes came under the ATF agency. The American Cancer Society was busy saying that diet had no effect on cancer, and they mentioned nothing about smoking and cancer.

As reports came out in different venues, then our supposed "watch-dogs" jumped on the bandwagon.

Currently our television and other media are virtually filled with pharmaceutical advertisements. Moreover, our "watch-dog" organizations are busy telling us to be careful of anything that isn't a pharmaceutical. Herbs aren't really effective; vitamins are not really helpful; homeopathy doesn't work; acupuncture is marginal and on and on. While at the same time, the FDA is approving drugs that have adverse effects such as coma, heart problems, breathing problems, stomach problems, stroke, diabetes, personality changes, blindness, thoughts of suicide and more. Am I the only person who is seeing that the side effects of our medical therapies are giving us some of the most horrendous health problems that we could possible have. This fact is astounding – astoundingly sad.

The Journal of the American Medical Association allows the companies to advertise to their doctors who read the journal about these wonderful products that so often DO HARM. I would challenge any person to look in the *Physician's Desk Reference* and find a pharmaceutical product that does not have a long list of adverse side effects.

As a researcher for almost 40 years, it has been a constant struggle for me to try to keep my composure and not go out into the world yelling and screaming for people to wake-up!

As I have gone around the world investigating, researching, documenting and compiling information on effective and safe therapies, technologies, clinics and practitioners, one thing has always come to my consciousness: Why aren't some of these things in active use today in the United States?

The reason is that they are not pharmaceutical products. The pharmaceutical industry has managed to lobby and get laws passed that keep pharmaceutical products as the legal products that can be used for various health problems. Now, why should this be possible? Why is it that new and possibly effective therapies can only be used once someone has gone through radiation and chemotherapy for cancer?

In the State of California the laws mandate nothing more than chemotherapy, radiation and surgery for cancer. All other forms of treatment are prohibited. To provide alternatives would make a physician a criminal!

How are we supposed to know if the new technique will work if the person has already gone through and failed some other therapy that has altered their body chemistry?

The answer is very simple. If the new tested therapy works despite the debilitated state the person is in, then the pharmaceutical company will say that it was really its product that worked. If the person dies or becomes worse, as chemotherapy or radiation might induce, then it was the fault of the newly used therapy.

I continually hear the statement from the medical establishment that if we waste our time on some alternative therapy, patients won't get the good and important care they need.

I've got news for the purveyor of this logic; the current techniques are not working for most of our problems. Despite the aggressive war on cancer for the last 40 years, death from cancer is still one of our main problems. We are talking of hundreds of billions of dollars spent on chemical research. In addition, the same is true with nearly every other illness and diagnosis.

A small but very important book was published in 1991 in Stuttgart, Germany. The 92 page document reports the results of an exhaustive statistical analysis of the world-wide literature (several thousand articles) on the use of cytotoxic chemotherapy in the treatment of advanced epithelial cancer.

Epithelial malignancies are responsible for more than 80% of cancer mortality in the world and include lung, breast, stomach, colon, pancreas, bladder, liver, ovary, cervix, trachea, bronchus, rectum, esophagus and head and neck.

The author of the book, Ulrich Abel, Ph.D., is a highly regarded cancer epidemiologist and biostatistician from the prestigious Heidelberg Tumor Institute. Abel based his conclusions not only on published material but on about 150 replies to personal inquiries he sent to more than 350 oncologists and research units around the world. The inquiries specifically addressed any unpublished data or any other data that would contradict his findings.

Abel's conclusion is stated on the first page in the preface: "There is no evidence for the majority of cancers that treatment with these drugs exerts any positive influence on survival or quality of life in patients with advanced disease." He went on to state, "However devastating the result of this work may appear, it is the undeniable conclusion of an attempt to evaluate the complete relevant literature by biometric standards and without professional bias."

Abel's devastating critique of chemotherapy was reported in an article in *Der Spiegel*, the German equivalent of our *Time* magazine.

The article observed that an English translation of Abel's work was made available at the Fifteenth International Cancer Congress held in Hamburg. More than 10,000 experts from around the world discussed and debated Abel's findings. In a follow-up article two issues later, Dieter Kurt Hossfeld, head of the Division of Oncology and Hematology of the Hamburg University Clinic in Eppendorf was interviewed. Hosssfoeld, a staunch chemotherapy advocate and chairman of the Organizing Committee of the congress, startled many physicians at the congress by his extremely critical comments about chemotherapy.

As I noted earlier, why would the State of California be legislating that this therapy be one of the only legal means of therapy?

We haven't discovered a cure for arthritis, diabetes, multiple sclerosis, ALS and many other problems. We mask the problems with chemicals and we never address the fundamental causes. We always hear that the answer is right around the corner, and with more financial support we will find the answer.

To give our support, we regular people run and walk for the cure, pray for the cure and plead for the cure to be found by our savior, the pharmaceutical industry.

However, we <u>have</u> <u>had</u> potential answers to the riddle of some of these illnesses and diseases. How is it that some of these techniques, despite being discovered or supported by some of the top medical doctors and scientists in the world, never made it into our medical system?

We found discoveries made by the Chairman of the Nobel Assembly, top research doctors from the Mayo Clinic, Johns Hopkins, Northwestern University, Yale University, University of Chicago and many other prestigious individuals from world-renowned universities, yet their work disappeared. Nothing happened with these wonderful discoveries and inventions.

At the time of this writing, we are in the midst of a presidential election. All of the candidates are addressing the issue crucial to so many Americans who don't have health insurance and access to health care. This is only part of a problem. What are we bringing these people into? What is our health care system doing to help people? We started the book with some facts regarding the state of health care. It is not a pretty picture.

Can we really say that the word "care" should be a part of the term *health care* when we speak about our medical system? I think not. It appears that you take your life in your own hands when you enter a hospital for any length of stay. When you take a drug you believe is intended to give you help and end up with a possible adverse effect that is equal to or worse than your existing problem, is this care?

I predict that we will also continue to receive reports regarding our pharmaceutical products having even greater side affects than has been stated when released to the medical doctors to be prescribed to patients.

Not too long ago I watched an interesting movie that was written by John Carpenter called, *They Live*. In the movie aliens had literally brainwashed the people of our planet to see only what they wanted us to see. Earthlings were put to sleep from thinking and saw what they were told to see.

If you do not have health, what do you have? You might be the prettiest woman in the world or the strongest man, the most powerful financial mogul, but if you don't have your health, what do you really possess?

I hope this book will initiate some important dialogue throughout our country. I have no doubt that the supporters of the drug approach will do what they can to defend themselves. Again, I must state that I realize that some pharmaceuticals have been beneficial. My objection is the heavy-handed manner in which the pharmaceutical companies have controlled legislation and eliminated all competition.

I would suggest you look at the health field like you do sports teams and their coaches. You measure the success of the coach and the team based on some type of performance level. What has really been accomplished? In sports, a coach might have led his team to an almost undefeated season, but if his lone loss is against a college rival, then the coach may be fired. We demand perfection, we demand success.

I believe that most people realize that the television commercials are truly stupid and ridiculous. There has not been one person I have spoken with who didn't roll his or her eyes and remark about the insanity of treating an ailment with a medicine that can cause dire consequences. But what are the alternatives? People don't question the status quo because they feel there are no other possibilities for finding answers. So we stick with what we have. I have presented only a few alternatives that come from a completely different approach. This approach is largely from the electromagnetic spectrum, which includes electricity, magnetism, sound, color, light and other frequencies.

In addition, I have mentioned one herbal approach that is a part of nature. All of the possibilities in this book are more natural and closer to the natural make-up of a human being than the chemical soup we are offered.

When we establish a medical system that is not based on greed but on truly finding answers, then we will find the health and happiness we all seek. I hope my words and thoughts have stimulated your interest. I don't want to think that I presented this book - and still nothing happened. **ONLY YOU CAN MAKE IT HAPPEN!**

146

What Must Happen
Chapter 16

We have so far discussed what has not happened. This is what I believe must happen to insure that we have healthier, happier and more productive lives.

- We must have an independent group of scientists and medical specialists review research in a variety of medical and health specialties. The structure would be similar to the concept of the Consumer's Union. The researchers would work in research centers that would include medical specialists, physicists and chemists.
- The pharmaceutical industry would not have any influence around or over the research centers, and no one who is or was employed within the pharmaceutical industry would be used as a consultant.
- Legislators could not determine what are or are not acceptable medical therapies. This is the province of the medical and health communities.
- Medical doctors should have the ability to use whatever therapy they deem best for their patients. This would include any and all therapies available from worldwide sources. Medical doctors' liabilities would be if they have deliberately harmed or did not use their best judgments for patients. After all, aren't medical doctors trained to find the best solutions for their patients. Why should someone else who is not a doctor legislate what doctors can use?
- Patients would have the right to decide what therapies would be available to them. If a therapy is not FDA approved then the patient would be informed and would sign a release. After all, patients already sign releases before surgical procedures and tests, so why shouldn't a patient sign to have access to some type of non-pharmaceutical or surgical procedure?
- People in the United States must have access to products and therapies that are used on patients with the same medical condition within other countries. If a product has been proved safe then it should be allowed to be used.

ONLY YOU Can Make It Happen!

If you are interested in learning more about the subjects contained in this book or would like to become more actively involved in freedom of medical choices, please visit our website:

www.andnothinghappened.com

You can reach the author at:

steve@andnothinghappened.com

Please tell at least three of your friends about this book. **"Free choice is not free; it must be pursued, defended and utilized."**

Appendix
Rife Microscope

    The following letters and newspaper articles were given to me by John Crane and were part of the collection of materials in the possession of Royal R. Rife.

    These materials have been part of lecture presentations and radio shows given by Steven Ross since 1983 and approved by John Crane. It is the author's belief that these materials have been in the public domain since that time.

EXECUTIVE DEPARTMENT
CITY OF SAN DIEGO, STATE OF CALIFORNIA
WALTER W. AUSTIN, MAYOR

December 1, 1931.

Mr. Royal Raymond Rife,
2500 Chatsworth Blvd.,
Point Loma, San Diego, Calif.

My dear Mr. Rife:

Permit me, as Mayor of
the City of San Diego, to extend
to you my sincere congratulations
upon the splendid work you have
accomplished along microscopic,
optical, bacteriological and mechanical
lines. Since you are young in
years you will have an opportunity
to do important work in the interest
of humanity and I wish you every
success. Should you find my office
necessary to any part of your work,
please call upon me.

Sincerely yours,

Walter W. Austin,
Mayor of the City of
San Diego, California.

```
                    ┌─────────────────────┐
                    │   ROYAL R. RIFE     │
                    └─────────────────────┘
          ┌──────────────────────────────────────────┐
          │        Milbank Johnson, M.D.             │
          └──────────────────────────────────────────┘
```

| Kendall, Ph.D. Northwestern Univ. School of Med. | O.C. Grunner, M.D. McGill Univ., Canada | Karl Meyer, M.D. Hooper Foundation Univ. of California. | Rufus von Klein Smid Pres. Univ of S. Calif. |

| E.C. Rosenow, M.D. Mayo Clinic | Dr. Walker, M.D. Univ. of California |

SPECIAL MEDICAL RESEARCH COMMITTEE
OF
THE UNIVERSITY OF SOUTHERN CALIFORNIA

Milbank Johnson, M.D. - Chairman

Rufus von Klein Smid, Ph.D. - U.S.C.

Whalen Morrison, M.D. - Chief Surgeon Santa Fe Railway

Charles Fischer, M.D. - Children's Hospital, New York

Karl Meyer, M.D. - Hooper Foundation, Univ. of California

A. Foord, M.D. - Pres. American Assoc. of Pathologists

A. Kendall, Ph.D. - Northwestern Univ. School of Medicine

Spencer Lens Company
Factory, Buffalo, N. Y., U. S. A.

LOS ANGELES, CALIFORNIA.

November 27, 1931

Mr. Roy R. Rife
712 Electric Bldg.
San Diego, Calif.

Dear Mr. Rife:

Just a short personal line to tell you that you have
made a very favorable impression on the scientific
people in and around Los Angeles. We recently heard
about a demonstration that you made at the California
Institute of Technology and many of my friends connected with the educational institutions have spoken
to me about the demonstration. It certainly has them
all "agog."

I also wish to extend to you my sincere thanks for
the very kind interview and time that you gave to a
very dear friend of mine, namely, Dr. Charles Chamberlain of the University of Chicago. Dr. Chamberlain
is well liked and loved by all who know him and you
have made an old man very, very happy.

With kindest personal regards, and assuring you of
my best wishes for your success, I am,

Yours sincerely,

Lyle D. Potter

LDP-GEA

Newest Microscope Will Trail Unknown Germs to Their Lairs

San Diego Inventor of Most Powerful Magnifying Instrument Known Tells Local Scientists of Marvelous Possibilities Now Opened to Research; Works on Still Greater Apparatus to Reveal Further Secrets.

By HAL JOHNSON

Has science, like the workman who neglects his tools, marched forward in the last 50 years without improving one of its most important complements, its optical equipment?

It has, according to Royal Raymond Rife, Ph.D., of this city, inventor of a powerful quartz-lense microscope, which has a magnifying power of 17,000 diameters and which may revolutionize the field of bacteriological research.

Speaking yesterday noon before a large gathering of prominent local physicians, scientific men and members of other professions at the University club luncheon, Dr. Rife contended that science has gone the limit with present microscopic and optical equipment.

"Science has gone ahead, but it has left its optical equipment behind," Dr. Rife said. "Lenses today are not so good as they were 50 years ago. Yet, in all great advances of scientific research one of the most important things was the optical equipment."

TO TRAIL STRANGE MICROBES

The next question of importance, which exists in the minds of laymen and men of science alike, is what use will be made of Dr. Rife's wonderful invention? Will it be put on the trail of heretofore obscure or unknown disease microbes and bacteria, to the great benefit of mankind in fighting pernicious maladies?

Dr. Rife said it will. He assured his listeners that he and several of his associates, men prominent in bacteriological circles, are going to work soon on a filterable form of infantile paralysis bacteria.

"If it is there," Dr. Rife said, "we can find it by breaking up the waves of light, so as to have no interference in the instrument.

"I've spent years looking down the tube of a microscope, but I have hardly scratched the surface. By the development of a more practical optical equipment, however, more details of a definite nature may be obtained with regard to ultra micro-organisms."

Combined efforts of Dr. Rife and Dr. Isaac Kendall, director of medical research at Northwestern university, may result for the first time in definite knowledge concerning bacteria which cause more than 50 communicable diseases.

FOLLOWS FILTERED GERMS

Dr. Kendall's discovery has to do with a substance, in which it is possible to cultivate germs as they develop in the human body. His process is one of filtering, but with continued filtering certain bacteria disappeared and it is Dr. Kendall's belief that the bacteria assume smaller forms. Excluding Dr. Rife's microscope, no instrument is said to exist which is powerful enough to see the filtered state. Commenting on the experiment yesterday, Dr. Rife said he predicted that the specimens would appear blue in the field. His prediction came true, he said.

One of the revolutionary features of the Rife microscope is that it makes specimens visible without use of stains. Dr. Rife said he believes micro-organisms are so chemically composed that they are susceptible to their own color stains. A system of rotating, wedge-shaped prisms in the Rife microscope, "bend" the light and aggregate one line of the spectrum. This is aided by a condensing and illuminating system of quartz lenses.

It apparently remains to create an index system of micro-organisms, according to their known or assumed colors. Dr. Rife said that each organism adopts its own individual band in the spectrum.

"We will know that the organism is not there," Dr. Rife said, "if there is no color.

"I positively believe the time is near when, for research work, we shall eliminate stain, and I believe a proper index can be worked out whereby each individual specimen will carry its own stain of the spectrum. It may take time, but other things in the field of scientific research have taken time."

USES GERM RANGE FINDER

In using his microscope at Pasadena, Dr. Rife said he also used a micro-polariscope, which is a microscope with a polarizer and analyzer attached. This aided in determining the chemical composition of the organism to be under observation. He explained that much time can be saved by use of the polariscope and a Vernier attachment, which aids in finding the object in the field. Without these finding devices, one might spend many hours hunting for the object, the speaker said.

Anyone who has gazed through the tube of an ordinary microscope can appreciate that statement. It sometimes is difficult to find the object in a microscope which magnifies 1800 diameters, or less. With an instrument of 17,000 diameters, the field would be like a chartless jungle.

Some idea of the magnifying ability of the Rife instrument is shown in experiments made at Pasadena, Dr. Rife said that his microscope even goes so far as to magnify vibrations together with the specimens, and that it becomes necessary for him and his associates to take the microscope to the basement and place it on the concrete floor to carry out their experiments.

The inventor announced yesterday that he is working on another microscope, which will embody a few changes. One of the biggest changes is in weight. The present instrument weighs 60 pounds. The new micro- science, but he pointed to an important feature of his instrument which differentiates it from the type microscope. In his microscope, Dr. Rife said, the rays or beams of light in the tube do not cross. They are held apart by six quartz lenses. Instead of crossing, the beams are brought to the full width of the tube and are condensed. He said that higher magnification has not been attained because of interference of light rays that would not allow amplification of the field.

Dr. Rife showed several microphotographs which he said he took at 17,000 diameters. This was made possible, he said, because of the remarkable revolving power of the instrument. This power, he said, was picked up in the photographs at a point where the eye failed to perceive. He said the revolving power of the instrument is three times greater than that of the human eye.

Dr. Rife said that parts for his microscope were made in many parts of the world. He said he used quartz glass entirely, because it allows from 48 to 50 percent more light than other kinds of glass. He said he uses a 3000-candle power illumination unit with the instrument. The beam of the light is cold, he said, and he has had a living specimen under it for five or six hours without evaporation from heat.

In introducing him, Gordon Gray,

WHAT'S NEW IN SCIENCE :: By RANSOME

IF THE experimental results obtained at the Pasadena Hospital by Dr. Arthur I. Kendall, bacteriologist of Northwestern University, and Dr. Royal R. Rife of San Diego test true, man now has the knowledge and a weapon which will enable him to win the war against disease-breeding germs, probably during the next decade. The story is almost unbelievable.

Dr. Kendall took a protein culture medium (his "Medium K"—prepared from the intestines of animals) and inoculated it with the well-known, rodlike typhoid bacilli. The bacilli of course multiplied rapidly. He then strained the culture through a Berkefield filter, thereby filtering out all the rods. Some of the culture medium, however, seeped through the invisible pores in the filter. These seepings were strained through two more sets of porcelain filters, and in the final seepings, bacteriologists have heretofore supposed, it would be impossible to find anything alive. But Dr. Kendall discovered that when these final filterings are placed in a peptone broth new bacilli form and grow to natural size, plainly visible under an ordinary microscope.

What is it that persists through three sets of filters, wherein the pores are too tiny for microscopes to detect, and starts life going again. "Seeds of life," the bacteriologist suspected, although the filtered liquid looked clearer than sterilized water. Nothing much larger than a molecule could have oozed through the three porcelain filters. Yet a drop of that liquid, when placed in a peptone culture, would bring back the rods that produce typhoid. It was getting close to the source of creation. For Dr. Kendall was able to say: "Let the waters bring forth abundantly the moving creature that hath life . . . whose seed is in itself," and the willing waters obeyed him.

Having heard about a "wonder microscope," said to have been invented by a young San Diegan, Dr. Kendall asked his friend, Dr. Milbank Johnson of Los Angeles, if such a microscope existed. Dr. Johnson did not know about it, but un-

Dr. Arthur I. Kendall, Northwestern University bacteriologist, left, and Dr. Royal R. Rife of San Diego with Dr. Rife's microscope.

THE WONDERWORK of 1931

demonstration—that the claims are possible. For there are factors which limit magnification. Lenses of high "resolving power" are too small to catch much light; hence, it has always been assumed, that what is gained in magnifying power may be lost because of deficient illumination. Immersion lenses with special curvatures, which bring the light to a focus in a drop of oil, increases the illumination, but that practice has its limitations. The absolute limit beyond which microscopes cannot go has been defined in these words: "Any object smaller than half the wave-length

Have No Fear, Says Foremost Authority, Man's Body Will Survive This 'Soft' Life

Dr. Lewellys Barker of Johns Hopkins Is Visitor Here

By MYRON V. DEPLW

THE "Sunday supplement scientists" who write weird bits intended to show how the human race is destined to become legless from too much dependence on automobiles, airplanes and street cars, are merely pulling the dear old public's pedal extremities.

Pleasant, alert, Dr. Lewellys Franklin Barker, one of the nation's foremost authorities on internal medicine, did not use any such undignified expression in discussing a typically "Sunday supplement science" question.

He sat erect in a chair in the lobby of Casa de Manana at La Jolla and considered the question seriously. He seemed to feel that, after all, the public might want to know whether such a thing were coming around.

The question:

"Is there any indication that the human body is undergoing changes to conform with the modern drug-store-lunch, got-to-hurry-along life? Are we in danger of becoming a race afflicted with accelerator toe or some such affliction because of the way we live?"

The answer of Dr. Barker was an emphatic no!

The emeritus medical professor of Johns Hopkins university, to whom Who's Who devotes half a column, had this to say:

STRAIN TOO GREAT

The nervous system attempts to adapt itself to new orders of things, to changing environments. Some of us are able to stand the speed and hurry, others are not and must seek less strenuous existence.

He agreed that the recent increases in nervous breakdowns, increasing suicides, have been due to nervous systems which have been unable to stand the shock of the economic depression.

But in the human body—the race—there has been no noticeable change so far as Dr. Barker has been able to observe.

"The effect of environment ... is still unmeas-

Dr. Barker

grow flabby and we may eventually lose the use of our legs. But that does not mean that our children will be born legless. There is nothing to indicate such will occur.

"Changes of that kind—race changes due to environment take a long, long time. It is something, as I have said, that still has not been measured."

Dr. Barker had just come from a visit with Dr. Raymond Royal Rife, inventor of a new microscope which recently created a stir in scientific circles when it was found the instrument produces a magnification of 17,000 diameters, nearly 10 times as powerful as others in use today.

"He said scientists were claiming the microscope was violating all the laws of optics and I told him we would just have to let them get some new laws then," Dr. Barker commented.

OPEN NEW FIELDS

He was obviously impressed with the instrument which Rife spent nearly 15 years perfecting. Through the new microscope, he said, science may hope to learn much about bacteriology, colloidal chemistry, and allied subjects, Dr. Barker said.

It will not, of course be of much assistance to institutions such as the Scripps Metabolic ... the chemistry of

Environment's Effect on Heredity Is Still Undetermined

only a week or so ago from Baltimore and will hurry back again Thursday of next week after three weeks spent in combining work with pleasure.

Beginning Thursday he will conduct a series of three clinics here.

Saturday evening he will be guest at a dinner given by Miss Ellen Scripps and J. C Harper, at which time the members of the county medical society will meet him. He will speak to them on the always popular subject of "Obesity."

VISITED RUSSIA

Plans are also being made for him to give an address in the Women's clubhouse Monday so that he may be able to tell of his findings in Soviet Russia. On his trip through Russia a year ago, Dr. Barker took a number of photographs which are used to illustrate the lecture.

Tuesday of next week Dr. and Mrs. Barker will leave with Dr. and Mrs. J. W. Sherrill for Imperial valley, thence to Los Angeles for a meeting with classmates and friends from Johns Hopkins. Then he will return east.

Dr. Sherrill, head of the Scripps Metabolic clinic, is arranging Dr. Barker's program during the visit here.

Dr. Barker is a native of Canada and received his early schooling at Pickering college. He attended also the Universities of Toronto, Leipzig, Munich and Berlin.

VARIED CAREER

In addition to being emeritus professor at Johns Hopkins, he is also visiting physician at Johns Hopkins hospital.

In 1899 he was Johns Hopkins medical commissioner to the Philippines, and in 1901 he was a member of a special commission appointed by the secretary of the treasury to determine the origin of the bubonic plague in San Francisco.

Dr. Barker is a member of medical societies both in this country and Europe and his books, papers and addresses, which are legion, command uni-

This Makes Things 31,000 Times as Big

Common Fly Grows Big as Rooster, Seen By New Microscope

Special to The Christian Science Monitor

SAN DIEGO, Calif.—A new, improved microscope, capable of magnifying without diffusion to 31,000 times, has been perfected here by Dr. Royal R. Rife, natural scientist, and has been on exhibition in the Bridges Fine Art Gallery, Balboa Park.

The new, shining instrument is a combination microscope and microspectroscope, and is expected to extend the boundaries of knowledge into new fields, since it is capable of ferreting out objects which hitherto have been invisible to human perception.

A little more than a year ago in his tiny laboratory above the garage on the estate of the late A. S. Bridges, his benefactor, on Point Loma, Dr. Rife invented a microscope with a magnification of 20,000 diameters. With it, Dr. Rife and Dr. Arthur L. Kendall of Chicago made some astounding discoveries in the realm of physics.

The new instrument, which Dr. Rife calls the universal microscope, was constructed on the same lines as the first, making use of the variable, wedge-shaped prisms which were the unusual feature of the Rife microscope.

Dr. Royal R. Rife's Micro-Spectroscope

THE CHRISTIAN SCIENCE MONITOR

Giant Microscope Explores New Worlds

REPORTED to be so powerful that it reveals disease organisms never seen before, the giant microscope pictured above has just been completed by Royal R. Rife, of San Diego, Calif., whose home-built instruments have long been ranked among the finest in the world. To eliminate distortion, the image produced by the new two-foot-tall apparatus does not pass through the usual air-filled tube, but along an optical path of quartz blocks and prisms. Weighing 200 pounds, the microscope has 5,682 parts.

157

SAN DIEGO UNION
, 1938   MAY 7, 1938

DREAD DISEASE GERMS KILLED BY RADIO WAVES, S. D. CLAIM

Specific Destroyer of All Deadly Microbes, Hope; Cancer Organism Isolated

By NEWELL JONES

Copyright, 1938, by The Evening Tribune

Discovery that disease organisms, including one occuring in dread cancer, can be killed by bombarding them with radio waves tuned to a particular length for each kind of organism, was claimed yesterday by a San Diego scientist, Royal Raymond Rife, Pt. Loma, according to a story by Newell Jones, copyrighted by The Evening Tribune and appearing in that newspaper yesterday. He added that he had isolated this cancer organism but is not positive yet that it is the direct cause of the disease.

The discovery promised fulfillment of man's age-old hope for a specific destroyer of all his infectious diseases, although Rife avoided any claim that he had established this yet. He announced his work in the conservative manner of scientists, but his reports indicated the great promise in their telling of successful bombardment of thousands of cultures of organisms, including almost all kinds known to afflict mankind.

Organisms from tuberculosis, cancer, sarcoma, the tumor resembling cancer but not so mortal as it; deadly streptococcus infection, typhoid fever, staphylococcus infection and two forms leprosy were among many which the scientist reported are killed by the waves. He said that his laboratory experiments indicated that the method could be used successfully and safely on organisms at work in living tissues.

GERMS 'DEVITALIZED'

"We do not wish at this time," Rife commented, "to claim that we have 'cured' cancer, or any other disease, for that matter. But we can say that these waves, or this 'ray,' as the frequencies might be called, have been shown to possess the power of devitalizing disease organisms, of 'killing' them, when tuned to an exact, particular wave length, or frequency, for each different organism. This applies to the organisms both in their free state and, with certain exceptions, when they are in living tissues."

The waves are generated in a new kind of frequency device developed by Rife and one of his associates, Philip Hoyland, Pasadena engineer. They are turned upon the organisms through a special directional antenna devised by the two.

Just what this Rife ray does to the organisms to devitalize them is not yet known. Because each organism requires a different wave length, it may be that whatever befalls these tiny slayers of man is something similar to the phenom-

Then, if he found one for that disease, he would have to start all over again on the next kind.

The scientist took first a culture of b. coli, the organisms which always seem to accompany the agency of typhoid fever yet apparently are harmless themselves. He prepared microscope slides from the culture and saw that his little subjects were alive. Then he turned the ray on them, tuned it to a certain frequency, then took the slide back to the microscope to see if anything

OBSERVATION CONFIRMED

The San Diego man explained that he found that different disease organisms have particular, individual chemical constituents and this led him to suspect that the organisms were electrical in character and might coordinate with variable electrical frequencies. His observations have been confirmed by certain British medical researchers, who say that they found that each kind of disease organism has a distinct radio wave length. So theoretically the Pt. Loma scientist's ray might make the tiny foes of mankind behave just as the vase and glass.

And, watched under the microscope, the organisms seem to do just that. When the ray is directed upon them, they are seen to behave very curiously; some kinds do literally disintegrate, and others writhe as if in agony and finally gather together in deathly unmoving clusters. Brief exposure to the tuned frequencies, Rife commented, brings the fatal reactions. In some organisms, it happens in seconds.

After the organisms have been bombarded, the laboratory reports show, they are dead. They have become devitalized—no longer exhibit life, do not propagate their kind and produce no disease when introduced into the bodies of experimental animals.

IDEA 18 YEARS OLD

The discovery of the ray's powers traces back, Rife recounted, to a day 18 years ago in his Pt. Loma laboratory. It was then his idea came to him. He has been tirelessly pursuing it to its conclusion down through all of those years.

Rife built a simple frequency generating apparatus of about 8 or 10 watts output. He grew some cultures of bacteria. Then he began the studies whose reported results now promise to revolutionize the entire theory and the whole treatment of human diseases, other than those of a functional or accidental nature.

Machine and cultures ready, the San Diegan anxiously, feverishly began testing his idea. Would those minute killers of men die under the frequency bombardment?

LONG, PATIENT WORK

It would be a patience-wracking task, for there was no way to measure what wave lengths, or frequencies, the organisms might have. In the quiet loneliness of the laboratory, Rife simply had to turn and turn and turn the tuning dials of his machine and, check, after each bombardment, the conditions of the

VIRUS HUNT SUCCEEDS

Inseparably linked with the ray development, Rife added, were two others almost equal in importance to the other discovery. These were a search for filter-passing viruses, those minute, disease-causing substances which sneak through the finest filters which scientists can to capture and study, and the designing and building of a microscope suited to the search, a microscope which would reveal to his eye viruses never seen before.

Both undertakings were successful, Rife commented. Eight years ago he began hunting the viruses with the microscope, and the satisfactory results, he said, made possible extension of the ray's use beyond the known disease organisms to others unseen and unknown before he ferreted them out.

Using a special media, or germ food, he prepared a culture from a human cancer. He gave the culture special treatment and incubation, he related. Finally it was ready, and he slipped a slide of it under his microscope, adjusted the instrument and anxiously fitted his gaze to the eyepiece. He saw on the slide a number of moving red-purple granules, the tiniest bit of microscopic life ever seen.

PRODUCES CANCER

And with those little, red-purple granules, Rife said, he produce typical, human cancer in rats!

The scientist frankly declared that he cannot be positive yet the tiny organisms are the direct cause of cancer. They have to b carried through three tests of experimental animals before they produce the cancerous tumors, he explained.

Appendix
Time Capsule Shorts

The majority of these short articles, appearing between 1930 and 1955, clearly demonstrate how electricity and magnetism exist wherever life is present.

Are we composed of electrical and magnetic particles and surrounded by *L-fields*? Can electricity and/or magnetism be used as a therapeutic agent to combat cancer, AIDS or other debilitating health problems? Based on the credibility of the researchers whose work is described throughout this book, you would certainly believe the fact that these areas in the world of physics demand more investigation.

The philosopher, Schoppenhauer, made the following statement: "Every new realization goes through three stages. The first stage, one laughs at a new realization. The second stage, the establishment will fight against it. The third stage, everybody wonders why didn't we have this a long time ago."

BIOLOGY

Electricity Exists Wherever There is Life, Scientists Find

If It Behaves in Living System as It Does Elsewhere
The Electrical Field May Determine Nervous Structure

By DRS. H. S. BURR and
PINCKNEY J. HARMAN, JR.
Yale University School of Medicine

(Electricity is the architect of the human body. Experiments by Drs. Burr and Harman, reported to the American Neurological Association, hold important implications in the understanding of health and disease in the human body, including perhaps even cancer.)

IT IS becoming increasingly clear that wherever there is life there is electricity. Apparently, a portion of the energy absorbed by a living thing from food and air and sun, is converted into electrical energy.

This energy is present in a relatively steady state just as in a battery there is a relatively steady voltage between the two poles of a battery. Some organs of the body, as for example, the heart and the brain, modify this direct current electricity to form an irregular alternating current which we recognize in the heart waves and in the brain waves. Studies of the direct current characteristic of the electricity found in living beings show that these are relatively stable but may be changed by fundamental biological activities such as menstruation, ovulation, cancer, growth, and wound healing.

However, it is well known that whenever electrical energy flows through a conductor, an electrical field can be found surrounding the conductor. Since electrical energy does flow through the living system, one should expect to find a field in that living system, unless electricity in living things is different from electricity in physical things.

May Play Major Role

If such a field can be demonstrated experimentally, it is by no manner of means impossible that it plays a major role in determining the pattern of organization in the living system. It has been possible to demonstrate the field experimentally by rotating a salamander on a revolving turntable under certain conditions. If this is done, the salamander produces a sine wave alternating current output analogous in every way except in frequency and magnitude of output, to that of the ordinary electric dynamo. This suggests that voltage gradients in the nervous system may be responsible for the presence in the nervous system of a field which determines the location of nerve cells and the direction of growth in nerve fibers.

Voltage gradients in the nervous system of the white rat have been determined in some forty animals and show that the brain is positive to the spinal cord and to the peripheral nerve. The voltage rises as anesthesia deepens, and lessens as anesthesia lightens. In no case is there any reversal of polarity. When the animal dies the voltage drops slowly to zero, usually within an hour. However, the voltage between the spinal cord and the nerve may persist for several hours.

Infer Nervous System Field

Since these voltages in the nervous system are analogous to those found in the whole living animal, and since the whole animal possesses an electrical field, it is logical to infer that the nervous system also possesses a field and it may well be that this field determines the structural arrangement of the parts of the nervous system.

Science News Letter, June 17, 1939

200,000,000 Electrical Particles in Every Breath

TWO hundred million particles in every breath a person exhales are the reason that the breath is visible on a very clear, cold morning.

Discovery of these particles, each nearly 100 times as big as an air molecule and which were previously unknown to science, was announced by Dr. George R. Wait, of the Department of Terrestrial Magnetism of the Carnegie Institution of Washington. The majority of the particles, he finds, are electrically charged, either positively or negatively.

Such particles, he said, are common in the air over chimneys, and the exhausts of automobiles. Perhaps those in the breath, he suggests, are the "smoke" of the fires of life itself, the constant burning in the body which keeps up its temperature.

On a cold morning, the moisture in the breath condenses around these particles. Consequently, it would be expected that in an open, snow-covered countryside, or in a desert, where the air is normally free of them the breath would be invisible. Even under such conditions, when the temperature is low, it can be seen, and this is explained by Dr. Wait's discovery of the particles in the breath itself.

The particles from the lungs, in a room where several people are assembled, quickly capture smaller ions, or broken air molecules, already present. He suggests that perhaps they play some part, as yet unknown, as carriers of disease-bearing microorganisms.

Science News Letter, May 24, 1941

Electricity, Not Beefsteak, To Prevent Black Eyes

➤ SOMETHING BETTER THAN the traditional beefsteak or hot or cold compresses for black eyes has been worked out by doctors at the Veterans Administration Hospital at Northport, L. I., N. Y.

It consists of a 20-minute treatment with a very small dose of electricity. Given within an hour or so after the injury, before much blackening of the eye has taken place, the treatment gave good results in more than 40 patients. When treatment was delayed until after marked blackening had developed, the results were not so good.

Use of this electrical treatment, called galvanism, is reported by Drs. Daniel Dancik and Anthony Degroot of the VA hospital in ARCHIVES OF PHYSICAL MEDICINE (Sept.)

Science News Letter, October 20, 195?

Eye Is Electric Generator, Current Tiny But Measurable

National Academy of Sciences, at University of North Carolina, Hears Reports From Many Fields of Research

"**E**LECTRIC glances," favorite phrase of old-time romantic novelists, is closer to literal fact than they ever guessed. For the human eye is an actual electrical generator, Prof. Walter R. Miles of Yale Medical School told the National Academy of Sciences, at the opening of their autumn meeting on the campus of the University of North Carolina.

The front part of the eye, Prof. Miles said, is electrically positive and the back part, where the retina is, has the opposite or negative charge. These differences in potential can be detected and measured by sticking thin pieces of metal foil on the skin at either side of the eye and attaching the wires to sufficiently delicate voltmeters. When the eye is held still, the instrument indicates steady voltage. As soon as you turn or roll your eye, you bring differently charged areas under the little electrodes, and the changes in the current show themselves on the dial.

Differences in potential, measured during a wide turning of the eyeball, range from .0002 to .003 volt for each eye. The amount of light falling on the eye at the time of measurement makes only a small difference in the result. One eye may differ markedly from its mate, just as people differ among themselves. Minor visual defects seem to make little difference.

That the eyeball itself, and not the surrounding muscle, is the source of the current was demonstrated when the tests were checked on persons who had lost one eye. If the eyeball is not there no current is generated, regardless of whether the socket is left empty or filled with a glass eye.

Cotton Under Microscope

COTTON, on which so much depends in the South and in the nation, was put under the ultra-microscope by Prof. Donald B. Anderson of North Carolina State College. The wall of the cotton fiber, he said, is composed of exceedingly small thread-like strands of cellulose that branch and interlace freely with each other, forming a close-meshed network.

In the oldest layer of the cell wall the cellulose strands lie in flat, close spirals, but in the parts formed later the strands wind in steep spirals. It is possible to control the rate at which the cellulose is laid down by controlling the environment. This may mean eventually a control of the quality and kind of cotton by suitable adjustments of the conditions under which it is grown, especially as regards temperature and light.

Increase Seed Size

DOUBLE the number of chromosomes in the cells of a plant and it will yield bigger seeds and show other changes in the direction of general giantism, Drs. A. F. Blakeslee and H. E. Warmke of the Carnegie Institution of Washington told the meeting. They provoked extra-chromosome plant types into existence by treating the parents with the drug colchicine.

Other effects of extra chromosomes Dr. Blakeslee mentioned were larger

MEDICINE

Electricity in Blood Clotting

Positive and negative electrical charges on blood chemicals may have part in blood clotting. Specially designed glass cell used in studies.

▶ POSITIVE AND negative electrical charges on molecules of chemicals in the blood may play a part in the mechanisms that make blood clot and keep it from clotting.

Studies on this new approach to the clotting problem were reported by Dr. Irving S. Wright of Cornell University, New York, at the meeting of the American Heart Association in Atlantic City, N. J.

Dr. Wright, like many other medical men, has for years worked on the problem of keeping blood from clotting in the blood vessels with serious consequences to the heart or brain.

He told of studies, using a specially designed glass cell, on the forces that tend to keep the fibrinogen molecules and other constituents of the blood separated and to keep them from sticking to the walls of the blood vessels and blocking them.

"Following the very fundamental law that two objects with similar electrical charges are mutually repulsive," Dr. Wright said, "it has been suggested that such mutual repulsion may play some part in preventing the components of the blood from coming together for clot formation."

"Our work has shown that, at least within certain limits, the anticoagulants (anticlotting drugs) tested increased the negative electrical charges, thus inhibiting the initial steps of clotting. However, much work must still be done before physical factors, including electrical charges of repulsion and attraction, can be conclusively established as playing a significant role in blood clotting."

Even now, with present methods, deaths and complications from clotting diseases can be reduced to a significant degree, Dr. Wright said. Among medicines doctors now can use in such conditions, he said, are heparin and dicumarol. Two newer anti-clotting drugs, tromexan and paritol, have not yet been used enough, in his opinion, for scientists to give them a final evaluation.

Science News Letter, June 16, 1951

MEDICINE

Polio Danger Period Month After Immunizing "Shots"

▶ BABIES AND children who need immunizing "shots" to protect them against diphtheria and whooping cough and perhaps other diseases should have these at least one month before the polio season starts.

This period, one month, is the time after the immunizing shots that a person is most likely to be in danger of getting paralyzing poliomyelitis, if exposed to the disease, the American Public Health Association warns today.

After one month "there is no evidence," the association states, that this effect of the shots persists.

Science News Letter, June 16, 1951

INVENTION

New Titanium Processing Methods Patented

▶ NEW METHODS for purifying and chlorinating titanium, important and strategic metal, used mostly in airplane metal alloys, received patents 2,555,361 and 2,—

Electricity of Blood Cells Enough to Light a Lamp

THE RED blood cells of man and animals as carriers of electricity are being studied at the Biological Laboratory, Cold Spring Harbor, Long Island, it was revealed before the meeting of the American Physical Society, in Rochester, N. Y.

Dr. Laurence S. Moyer and Dr. Harold A. Abramson reported that red blood cells of man, among the animals studied, have the highest effective electrical charge at their surface, equivalent to 15,000,000 electrons. Electrons are the unit charges of electricity.

Studies of the amount of electricity carried by the blood cells have an important relationship to such basic human problems as the coagulation properties of the blood and problems connected with the anemias. For example, it has been found that in certain cases of anemia in human beings, the abnormal cells apparently possess a mechanism which is capable of preserving the normal surface charge of the cell in spite of wide variations in the surface area during the course of the disease.

A good idea of the size of this surface charge may be obtained from the estimation that if the charges from blood of a full-grown man could be collected and made to pass through a 25-watt electric bulb it would burn for at least 5 minutes.

Of all the animals studied in the tests Drs. Moyer and Abramson found that the rabbit had the lowest electric charge density—only 1,890 electrostatic units. Man and the rhesus monkey (used in experimental studies of infantile paralysis) had about the same charge density,

4,500 units. The dog had the highest charge density, 5,600 electrostatic units.

Electrons Born

Sprays of electrified particles shoot out, now and then, from all kinds of matter. Rocks, metals, even our own bodies, are subject to this effect which physicists say is due to the unceasing cosmic-ray bombardment.

Disruption of atomic nuclei by the highly energetic cosmic-ray particles has been regarded as a likely explanation. The particles making up the spray were thought of as the flying debris from a shattered atom.

But it now seems more likely that the atoms remain intact during the collision and that the cosmic rays suffer the major damage. According to Dr. and Mrs. Carol G. Montgomery of the Bartol Research Foundation of the Franklin Institute the spray particles are pairs of positive and negative electrons created in that intense electric field which surrounds the nucleus of every atom. The raw material for the process is the energy of the cosmic-ray photons.

Dr. Montgomery described to the meeting experiments which he and his wife performed with a device called an "ionization chamber." Different kinds of material—lead, tin, iron, magnesium—were piled about the chamber and their electrical effects recorded on yards and yards of photographic film.

The heavier the material surrounding their chamber, the larger was the number of particles shot out in every spray. Heavier atoms have stronger electric fields about them; have greater power they could support on a continuous-yield basis in times of full rain. And during the present dry half of the irregularly recurring climatic cycle, those mouths, desperate with starvation, have skinned the grass down to the very roots. There is nothing left to eat, nothing left to burn.

Perhaps there are plainsmen who still remember old times with something of a pang—who would not be wholly sorry to see an old-fashioned prairie fire again, because if it did symbolize destruction, it also symbolized a high abundance even as did the "whole burnt offerings" of ancient Israel.

Science News Letter, July 18, 1936

PHYSIOLOGY

Electrical Changes in Body Controlling Factor in Growth

By PROF. H. S. BURR, Yale University School of Medicine

IN ALL probability, wherever there is life, electrical phenomena are to be found. Electrical studies of living plants and animals have added much to our information since Galvani first published his description of "Animal Electricity."

Great progress has been made in the study of the nervous system through the adaptation of recent commercial radio apparatus to this use. However, it has been very difficult to determine with precision the nature of the electrical currents which have been noted in association with living animals and plants since most of the meters used require current for their operation, and hence have complicated the results by the effects of changes in resistance.

To overcome these difficulties, a vacuum tube microvoltmeter has been developed which is stable, draws no current and is, therefore, independent of resistance. With this instrument, differences as small as a millionth of a volt can be read accurately. Reproducible voltage differences of a characteristic order in fishes, salamanders, frogs, chicks, rats and mice, cats, rabbits, dogs, monkeys and man have been obtained.

Accompanying Life Processes

Moreover, it has been shown that these voltage differences are very closely associated with minute variations in the living process. The instant of ovulation in the intact cat and rabbit and an electrical rhythm in the menstrual cycle in women has been recorded. A marked change associated with the appearance of cancer and a definite correlation with growth during the embryonic and adult life have been observed.

In a rather surprising way it has been found that the voltages developed are not the result of chaotic currents but of currents organized into a very definite pattern which is characteristic of the species and may show the same individual differences as do series of finger prints. With this instrument, it is possible to write a kind of electrical formula for the individual animal.

"Prospect" the Body

In addition, it is possible to study electrically a live animal with very great accuracy without having to kill it for analysis or without in any real sense modifying its activities. In fact, it is possible to prospect the body of an animal for voltage differences much as a geophysicist maps the surface of the ground for hidden ore. Plots constructed in this way give numerous clues to what is going on inside the animal.

It has been found, moreover, that readings taken from any two points on the body reflect not only what is going on in the immediate vicinity of those points but also the *total* activity of the animal. Every animal so far studied produces electricity in amounts that can be accurately measured.

The data suggest that each animal possesses a dynamic electrical picture which, although constantly changing in minor ways, nevertheless, possesses recognizable individual characteristics.

There is a very real possibility that this electrical picture or electrodynamic field may provide the explanation of the amazing capacity of an animal to grow from a single egg into a multiple-celled adult in the midst of the rapidly changing chemistry of development.

Hope For New Clue

It may be that in these electrical studies will be found the clue to the mechanisms by means of which the chromosomes determine such things as shape of face and color of eyes, and that "animal electricity" is the expression of a fundamental electrical field acting as a guiding and controlling factor in the development of any individual.

Science News Letter, July 18, 1936

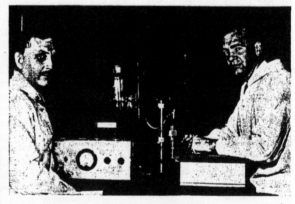

STUDY BODY CURRENTS

How Yale experimenters study electricity of nervous system by adapting radio apparatus to this purpose. Prof. H. S. Burr, left, is ready to make measurements upon Dr. R. G. Meader, right, who has his fingers in salt dishes to make electrical contact between his body and the sensitive instruments. Taking of readings is accomplished simply and without discomfort to the person being tested.

Science News Letter, July 18, 1936

Electric Current of Cells Expels Body Fluids

DELVING into the choroid plexus, which is the part of the brain where the spinal fluid is manufactured, Drs. Louis B. Flexner and Robert D. Stiehler of the Johns Hopkins Medical School have discovered that an electric current drives this fluid out of the cell manufacturing plant and propels it into the spinal canal.

While their research applies only to the choroid plexus and the spinal fluid, it is a beginning attack on the unsolved problem of how other substances manufactured by the body cells, hormones for example, are pushed out into the body.

The electric current which does this for the spinal fluid is generated by the energy developed by the cells of the choroid plexus as they use oxygen to burn food for nourishment.

An important feature of this research was the discovery that a complete electric circuit is formed in the body. Previously scientists believed that the circuit could not be complete until they added an electric wire. The Johns Hopkins scientists found that in the choroid plexus the circuit is completed by a membrane which can carry electric current like a wire. This membrane lies between the connective tissue and the outer covering of the structure.

Science News Letter, April 16, 1938

when patent number 2,231,929 was granted to Joseph Lyman for a radio plane locater. The patent was assigned to Sperry Gyroscope Company, of Brooklyn, N. Y., which makes control mechanisms for airplanes and ships. (See SNL, March 8.)

Science News Letter, June ... 1941

Life and Death Closely Connected With Electricity

Live Tissue Is Like B-Battery and Dead Tissue Like Burned-Out Generator, Engineers Are Told

LIFE and death are very closely connected with electrical activity, Dr. Robert S. Schwab, of the Brain Wave Laboratory of the Massachusetts General Hospital and the Harvard Medical School, told the American Institute of Electrical Engineers.

"In these days of super-sensitive amplifiers and recording apparatus," he said, "it is very tempting to define life and death in terms of electrical activity. Whether or not this concept is accurate, we can, on present knowledge, liken living tissue to a B-battery and dead tissue to a burned-out generator. The function of living tissue, however, is so closely allied with its electrical activity that knowledge of the latter has given us better understanding of the working of the human body."

There are four types of body electrical currents, he stated. One is a small direct current in which the cells act as a B-battery. Second is an alternating current wave that accompanies contraction of muscle tissue. It also occurs in connection with activity of nerve fibers. Third is the type "associated with the more highly developed types of contractile tissue," such as the heart. This is used to operate the electrocardiograph, important instrument enabling physicians to diagnose heart ills.

"The fourth type of body electricity," said Dr. Schwab, "is that associated with the complicated tissue that makes up the central nervous system of animals. Here, as the function is continuous during life, the ganglia and brain cell tissues are ever-active electrically and show no periods of rest in the manner of muscle and nerve. Each brain cell does not actually beat alone, but by a system of interconnections they keep each other stimulated to activity. These 'chains' ... make up the bulk of the

brain and spinal cord of man and animals."

The number of possible combinations of neurones, he stated, is represented by the number 1 followed by 2,783,000 zeros, which is greater by far than the number of electrons and other elementary particles in the entire universe, according to astronomical estimates.

These currents require extremely delicate recording equipment, but they show waves of different kinds which have been very useful to physiologists in studying the cells of the central nervous system.

Science News Letter, June 28, 1941

INVENTION

Windshield Protected From Insect Attack

A TRIANGULAR screen of transparent plastic, mounted at the front of an automobile above the radiator grille, diverts currents of air around the sides and above the car. Insects are carried with the air, so they cannot soil the windshield and cause possible danger from obstructed vision. (Henry Mfg. Co., Minneapolis.)

Science News Letter, June 28, 1941

GEOGRAPHY

Emperor Penguins Pictured On Antarctic Expedition

See Front Cover

NAVY photographer Charles C. Shirley of San Diego, Calif., waited six hours on the Bay of Whales in the Antarctic dawn temperature of 20 degrees below zero to secure the beautiful photograph shown on the front cover of this week's SCIENCE NEWS LETTER.

This photograph of the Emperor Penguins is an official photograph of the U. S. Antarctic Service.

Science News Letter, June 28, 1941

POLAR AIRDROME

Like lump sugar are these white building blocks of snow with which a working party are building a winter hangar for the small cabin plane at West Base. The picture, taken at 68 degrees below zero, is an official photograph of the U. S. Antarctic Service.

Science News Letter, June 28, 1941

Electricity of Human Cells

When sitting or lying still you burn energy at the rate of a 100-watt lamp by passing electrons over compound "batteries", connected in series.

▶ YOUR BODY uses up energy at about the same rate as a hundred-watt lamp when you are sitting or lying still, Prof. Eric G. Ball of Harvard Medical School stated before the meeting of the American Philosophical Society in Philadelphia. Like the lamp, the body obtains this energy by a process which involves the flow of an electric current.

"In the living cell, electrons flow from the foodstuffs we ingest to oxygen, thus reducing the oxygen to form water," he continued. "The 'filament' of the cell over which these electrons flow is not of uniform composition as it is in a light bulb. The electrons in the cell are passed along over a chain of compounds composed of iron-containing proteins, the cytochromes, and vitamin-containing units named coenzymes.

"The over-all process involves a potential change of about 1.17 volts and a total flow of current in all the body cells which amounts to about 76 amperes. The process occurs, however, in a step-wise fashion which involves five or six successive transfers of electrons between the various components comprising the cellular 'filament' or oxidative chain. Each pair of components may thus be looked upon as forming a battery, with the pairs connected in series. A drop in voltage occurs with the interaction of each pair in this series, the magnitude of which may be estimated from our knowledge of the oxidation-reduction potentials of each of the systems involved."

Science News Letter, May 3, 1947

Heavy Oxygen Will Aid Studies in Chemistry

▶ A KIND of heavy oxygen, with an atomic weight of 18 instead of the usual 16, can now be used to settle long-disputed points in chemistry and physiology, Dean Hugh S. Taylor of Princeton University told the American Philosophical Society. The isotope-separation techniques developed by the Manhattan District make this type of oxygen available for research purposes in any reasonable quantity if the cost can be met.

As an example of the long-standing problem already solved with molecules "tagged" with heavy oxygen, Dean Taylor mentioned the fates of water and carbon dioxide taken in by plants. Both compounds contain oxygen, the sum of which is in excess of the plant's needs for its food- and body-building processes. Plants have long been known to give off oxygen: where did it come from? By the use of "tagged" molecules of water and carbon dioxide it has now been demonstrated that the oxygen going in with the carbon dioxide stays in as part of the plant structure, whereas the oxygen that goes in as part of water comes out again as pure oxygen.

Science News Letter, May 3, 1947

PLANT PHYSIOLOGY

Restoring Sight, Sound

Based on nerve research now in progress, prediction was made that stimulating the brain electrically would restore sight, hearing and motion to the disabled.

➤ MAKING the lame walk, the blind see and the deaf hear will come out of the realm of miracles and into everyday life when brain and nerve research now under way reaches its goal.

Some of the research which, in theory, makes possible this prediction was reported by Prof. Wendell J. S. Krieg of Northwestern University Medical School at a meeting at the University's campus in Evanston, Ill.

Electrodes placed in or on the brain, to stimulate by electrical impulse the motor points of muscles or the seeing and hearing points of the brain, are the means which Prof. Krieg believes will accomplish the miracles.

A person who has become blind may again obtain the sensation of light if a point at the back of the brain is stimulated electrically, he said. If a number of electrodes were distributed over the surface of the brain in a prearranged pattern or sequence, the person might perceive outline or movement. He could read if single letters were transmitted one at a time as in news flash signs by trainer operating a typewriter whose keys set off electrical switch patterns.

"It is only a technological step," Prof. Krieg said, "to devise an appliance to scan the visual field in the same manner as a television scanner and to transmit that which is seen and recorded to the cortex (of the brain) in the same sequence and scanning pattern."

The fact that man sees and hears with his brain, not with his eyes and ears, is what makes such applications possible. Eyes and ears merely receive and transmit stimuli. Use of electrodes to provide the stimulating signals will become possible, Prof. Krieg believes, as scientists learn more about the brain and nerve tracts and which ones to stimulate for sight, hearing and muscle movement.

Although much basic study will be needed before the theory can be put into practice, it "is not so much in the clouds as it sounds," Dr. Richard H. Young, dean of Northwestern's medical school, commented.

Future progress and research in neurology will be in this direction, he said, and achieving the goal of restoring sight, hearing and muscle movement is "perfectly possible."

Science News Letter, November 19, 1949

Old Brain Drives the Intellect Via Feelings

➤ THE old brain, old in the sense that it came first in man's evolutionary development, deserves more credit than it usually gets, Dr. Stanley Cobb, Harvard University neurologist and psychiatrist, declared at the New York Academy of Medicine in New York.

This old brain is the part of the brain through which we feel and smell and it is much more than a relic left over from an earlier evolutionary stage for the lost art of living by smell.

"To the sorrow of many persons who believe they can rule their emotions by intellectual will power," Dr. Cobb said, this

Simple Electrical Model of a Nominal Biological Cell

The Cell theory tells us that the cell is the basic unit of life, all organisms are made up of cells, and new cells come from other living cells. Whether you are a simple organism, an amoeba, made up of only one cell or a complex organism, a human, made up of many cells, each cell is built the same way.

It should become obvious from this picture that the cells of our body our going to be influenced by electromagnetic fields of the body.

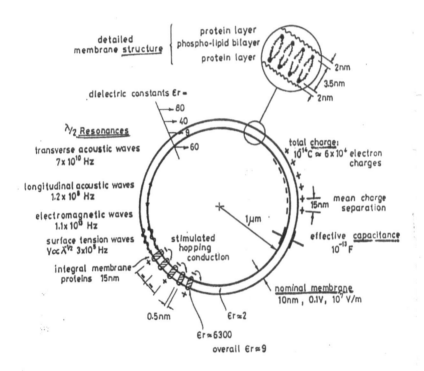

Fig. 1 Simple electrical model of nominal biological cell.

Suggested Reading

Chapter 2 Surpassing All Limitations

Bird, Christopher, *What Has Become of the Rife Microscope*, New Age Journal, March, 1976.
Gallert, Mark, *New Light on Therapeutic Energies*, James Clarke & Co, 1966.
Lynes, Barry, *The Cancer Cure That Worked*, Compcare Publications, 1987
Seidel, R.E. and Winter, Elizabeth, *The New Microscopes*, Annual Report of the Smithsonian Institute for 1944.

Chapter 3 The Nobel Chairman Is Ignored

Nordenstrom, Bjorn, *Biologically Closed Electric Circuits*, Nordic Medical Publications, Stockholm, 1983
Nordenstrom, Bjorn, *Biologically Closed Electric Circuits*, DVD presentation of lecture and seminar, World Research Foundation Congress of 1986, World Research Foundation, Sedona, Arizona, 1986, www.wrf.org

Chapter 4 Who Is Colorblind?

Clark, Linda, *The Ancient Art of Color Healing*, Pocket Books, 1978.
Dinshah, Darius, *Let There Be Light*, Dinshah Health Society, Malaga, New Jersey, 9th Ed.

Chapter 5 Waves That Heal

Brown, Thomas J., The Lakhovsky Multiple Wave Oscillator Handbook, Borderland Sciences, 1994
Clement, Mark, *The Waves That Heal*, Health Research, 1963
Lakhovsky, Georges, *The Secret of Life*, True Health Publishing Company, 1963

Chapter 6 Fields of Life

Burr, Harold Saxton, *Blueprint for Immortality*, Redwood Burn Limited, 1982
Ravitz, Leonard, *Electrodynamic Man*, Rutledge Books, 2002
Russell, Edward, *Design For Destiny*, Biddles Ltd., 1978

Chapter 7 The Phenomenon of Life

Crile, George, *A Bipolar Theory of the Living Processes*, The Macmillian Company, 1926
Crile, George, *The Phenomena of Life*, W.W. Norton & Co., 1936

Darras, Jean-Claude and DeVernejoul, Pierre, *Visualization of Acupuncture Meridians*, World Research Foundation Congress 1986, DVD, World Research Foundation, Sedona, AZ., www.wrf.org.

Additional Reading
Technical and Non-Technical

Becker, Robert O. and Selden, Gary, *The Body Electric*, William Morrow Co., 1985

Becker, Robert O., *Cross Currents*, Jeremy P. Tarcher, 1990

Brodeur, Paul, *Currents of Death*, Simon and Schuster, 1989

Capra, J., *The Tao of Physics*, Bantam Books, 1977

Colson, Thomas, *Molecular Radiations*, Electronic Medical Foundation, 1953

Dale, Cyndi, *The Subtle Body*, Sounds True, 2009

David, W., *The Harmonics of Sound, Color & Vibration*, Devorss & Co., 1980

Davis, Albert and Rawls, Walter, *Magnetism and Its Effects on Living Systems*, Exposition Press, 1974

Davis, Albert and Rawls, Walter, *The Magnetic Effect*, Exposition Press, 1977

Dossey, L., *Space, Time, and Medicine*, Shambhala, 1982

Dumitrescu, I. and Kenyon, J., *Electrograpic Imaging in Medicine and Biology*, Neville Spearman, Ltd., 1983

Eden, James, *Energetic Healing*, Insight Books, 1993

Jenny, Hans, *Kymatic Cymatics*, Basler Druck, 1967

Le Bon, G., *The Evolution of Forces*, D. Appleton and Co., 1908

Le Bon, G., *The Evolution of Matter*, Walter Scott Publishing Co, 1909

Gallert, M., *New Light on Therapeutic Energies*, James Clarke & Co, 1966

Gerber, R., *Vibrational Medicine*, Bear & Co., 1988

Moss, T., *The Body Electric*, J.P. Tarcher, Inc., 1979

Popp, Fritz-Alpert, *Electromagnetic Bio-Information*, Urban & Schwarzenberg, 1979

Rahn, O., *Invisible Radiations of Organisms*, Verlag von Gebruder, 1936

Talbot, M., *The Holographic Universe*, Harper Collins, 1991

Tiller, W., "*Energy Fields and the Human Body*", *Frontiers of Consciousness*, White, J., Editor, Avon Books, 1974

For therapeutic information, concerning specific health problems contact:

World Research Foundation
41 Bell Rock Plaza
Sedona, AZ 86351 USA
www.wrf.org